Poverty, Food Consumption, and Economic Development

Maneka Jayasinghe

Poverty, Food Consumption, and Economic Development

Springer

Maneka Jayasinghe
Asia Pacific College of Business and Law
Charles Darwin University
Darwin, NT, Australia

ISBN 978-981-16-8742-6 ISBN 978-981-16-8743-3 (eBook)
https://doi.org/10.1007/978-981-16-8743-3

This Springer imprint is published by the registered company Springer Nature Singapore Pte Ltd.
The registered company address is: 152 Beach Road, #21-01/04 Gateway East, Singapore 189721, Singapore

To Amma for the immense sacrifices you made for me

and

to Mamma for inspiring me to become an Economist.

Acknowledgements

I wish to express my sincere gratitude to my mentors Prof. E. A. Selvanathan and Prof. Saroja Selvanathan at Griffith University for their encouragement and guidance to undertake research. I gratefully acknowledge the constructive comments on various chapters and encouragement from Emeritus Professor Ken Clements at The University of Western Australia. I also acknowledge the comments and suggestions of the anonymous reviewers and the editor of the book at *Springer Nature (Singapore) Publishing Limited*. I would also like to thank Emeritus Professor Christine Smith, Associate Professor Andreas Chai, and Dr. Shyama Ratnasiri at Griffith University for helpful comments on various chapters of the book. I wish to thank Ms. Tanya Parker, Ms. Saranya Balasubramanian, and Mr. Umamagesh Perumal for their excellent editorial assistance and Ms. Anushangi Weerakoon for her incredible support to publish this book.

I gratefully acknowledge the research grant received from the College of Asia Pacific Business and Law, Charles Darwin University to complete this book. My special thanks are also extended to the Department of Census and Statistics, Sri Lanka for giving me permission and access to the Household Income and Expenditure Survey data.

Last but not least, my heartfelt appreciation is extended to my family and friends for their encouragement, support, and understanding.

Acknowledgements

Technical Notes

This book contains seven chapters. To aid the reader, each chapter has been written so that it is largely self-contained.

Each chapter contains several sections, subsections, and a list of references. The sections in each chapter are numbered at two levels: the first level refers to the chapter and the second level refers to the section in each chapter. For example, Sect. 2.3 refers to the third section in Chap. 2. The subsections are numbered at three levels: the first level refers to the chapter, the second level refers to the section, the third level refers to the subsection, and fourth level refers to the sub-subsection. For example, 2.3.3.1 refers to the sub-subsection 1 in the third subsection in third section in Chap. 2.

Equations are indicated by two numbers: the first refers to the chapter and second to the order of the occurrence of equations within that chapter. For example, Eq. 2.3 refers to the third equation in Chap. 2.

Tables and Figures are indicated by two numbers: the first refers to the chapter and the second to the order of the occurrence of tables or figures within that chapter. For example, Table 3.2 refers to the second table in Chap. 3 and Fig. 3.1 refers to the first figure in Chap. 3.

Technical Notes

Contents

About the Author

Maneka Jayasinghe is a Senior Lecturer in Economics and Head of Business in the Asia Pacific College of Business and Law at Charles Darwin University, Darwin, Australia. She has also taught previously at Griffith University, Queensland, Australia. Prior to joining academia, she has worked at GTZ and UNICEF in Sri Lanka. She holds a Ph.D. in Economics (Griffith University, Australia), Masters in Public Policy (National University of Singapore, Singapore), and Bachelor of Arts, Economics Honours (University of Colombo, Sri Lanka). She is an Applied Economist and her research interests include Consumer Demand Analysis, Poverty and Wellbeing, Energy Economics and Tourism Analysis. She has published numerous articles in leading international refereed journals such as *Energy Economics*, *Journal of Travel Research*, *Applied Economics*, *Economic Record*, *Social Indicators Research*, *International Journal of Tourism Research*, and a number of book chapters with leading publishers.

Abbreviations

CCPI	Colombo Consumer Price Index
DCS	Department of Census and Statistics
EPEA	Expenditure Per Equivalent Adult
FAO	Food and Agricultural Organization
FGTI	Foster Greer-Thorbecke Index
FHHs	Female-headed Households
HI	Headcount Index
HIES	Household Income and Expenditure Survey
IV	Instrumental Variable
LTTE	Liberation Tigers of Tamil Eelam
MHHs	Male-headed Households
MRS	Marginal Rate of Substitution
OECD	Organisation for Economic Cooperation and Development
OLS	Ordinary Least Square
OPL	Official Poverty Line
PCE	Per Capita Expenditure
PCI	Per Capita Income
PGI	Poverty Gap Index
SDG	Sustainable Development Goals
SLR	Sri Lankan Rupees
SPGI	Squared Poverty Gap Index
SPI	Spatial Price Index
SUR	Seemingly Unrelated Regression
UNDP	United Nations Development Program

List of Figures

List of Tables

Chapter 1
Introduction

1.1 An Overview

There has been a notable decline in poverty globally, with the reported number of people living in extreme poverty having declined from 36% in 1990 to 10% in 2015. However, new challenges such as COVID-19 hinder global efforts to eradicate poverty. According to the World Institute for Development Economics Research at the United Nations University, the economic downturn resulting from the global pandemic could increase global poverty by about 8% of the total population, increasing the poverty levels for the first time since 1990.

In order to tackle poverty efficiently, it is of utmost importance that governments and international aid organisations make accurate assessments of household standards of living and welfare. A crucial step in such poverty assessments for any country is the comparison of welfare across households with different sizes and compositions and in different sub-national locations. As far back as 1895, scholars such as Ernst Engel realised that one cannot simply compare incomes across households of different sizes and compositions. This is because larger households tend to be better off than smaller households if the economies of scale in consumption associated with large families are taken into consideration when assessing poverty. For example, a household with two members would certainly need to consume more than a single-person household. However, a household with two members does not require two times as much electricity, durable goods, housing space and even food as the single-person household if the former wants to enjoy an identical level of living standard of a single-person household.

This phenomenon can be pivotal to the development of targeted strategies to alleviate poverty in particular segments of the population. The most common way to measure economies of scale in consumption and thereby adjust household income to incorporate economies of scale into poverty measurements is the use of equivalence scales. Nevertheless, a practical issue relating to the use of equivalence scales arises from the complexity in estimating equivalence scales and the lack of consensus on

M. Jayasinghe, *Poverty, Food Consumption, and Economic Development*,
https://doi.org/10.1007/978-981-16-8743-3_1

the best method to estimate equivalence scales. To circumvent this problem, policymakers tend to use a simple form of equivalence scales, such as the Organisation for Economic Cooperation and Development (OECD) modified equivalence scales, to adjust household expenditure in order to compare the welfare of households with different sizes and compositions.

An important drawback of using these predefined equivalence scales is that they may not provide an accurate assessment of the magnitude of economies of scale in consumption in a particular context. This is because the level of consumption economies of scale (and hence the equivalence scales) can be affected by various socio-economic, demographic and behavioural choices of households, which are not identified in those simple equivalence scales. If the OECD method is not suitable, what is the most appropriate method to construct more nuanced equivalence scales? As discussed in Chap. 2, there is extensive academic literature on approaches to calculating various equivalence scales, such as the Barten method, the Rothbarth method, and the Engel method. The attractiveness of the Engel method lies mainly in its definition of welfare as the percentage of total household expenditure spent on food consumption. This is particularly important because, in developing countries, a major share of household expenditure is spent on food consumption. The Engle method also enables the incorporation of various socio-economic, demographic and behavioural choices of households into equivalence scale calculations.

While there is a considerable body of literature on economies of scale in consumption and their implications on poverty measurements, several knowledge gaps in the literature have been identified. Firstly, there is a dearth of empirical evidence on the economies of scale in the consumption of home-grown food. This exclusion in the literature is a serious issue, as home-grown food fulfils a major share of the food and nutrition needs of rural communities and thereby makes a substantial contribution to food security in developing countries. Subsistence farming also supplements the monetary income of these communities to a considerable extent. There exist significant variations in home-grown food consumption behaviours among population sub-groups living in rural and urban areas. Therefore, understanding economies of scale in home-grown food consumption is of utmost importance when comparing the standard of living of population sub-groups in developing countries.

Secondly, despite evidence of significant changes in relation to food consumption behaviour, advancing domestic technology, and widespread use of efficient energy sources facilitated by household electricity access, particularly in developing countries, the role of these factors in determining the level of economies of scale in food consumption has received limited attention so far. Such an omission in the literature is unfortunate, as the food consumption behaviour of the majority of the population in developing countries is undergoing significant changes due to massive socio-economic advancements having occurred in recent years. For example, increasing household incomes, increasing access to electricity and on-going urbanisation may have resulted in a considerable increase in household usage of domestic appliances in food preparation and storage compared with the past, which may, in turn, have a noteworthy impact on the realisation of economies of scale in food consumption.

Thirdly, the link between economies of scale in consumption and gender remains largely unexplored. The idea that female and female-headed households disproportionately represent the poor, and that they experience a higher incidence of poverty than male and male-headed households, has been discussed widely in the literature on gender and poverty, particularly in the context of developing countries. Nevertheless, important questions remain unexplored. For example, how does the gender of the headship affect household ability to reap the benefits of consumption economies of scale? And how do these differences in consumption economies of scale, based on the gender of the household head, change the poverty status in male-headed and female-headed households?

The central research question of this research is: *How do the households' characteristics and behavioural choices affect the level of food consumption economies of scale enjoyed by households and what are the implications of these food consumption economies of scale on household poverty status?*

In the process of answering this central research question, this book specified four additional research questions that have both academic and policy relevance. Specifically,

R.Q.1. How does the household choice to consume home-grown food affect food consumption economies of scale and enhance food security?

R.Q.2. How does domestic technology adoption and household electricity access affect food consumption economies of scale?

R.Q.3. Do food consumption economies of scale vary by the gender of the head of the household?

R.Q.4. How do the food consumption economies of scale affect the poverty measurements?

A number of quantitative research methods, such as Engel equivalence scales to measure economies of scale and regression analysis have been used in this book to address these research questions. Additionally, two theoretical models have been proposed and empirically tested to explore and validate two research questions (R.Q.1 and R.Q.2).

This research is expected to provide some important insights for policy development aimed at improving the standard of living of households. Using a series of interrelated studies, this book highlights the role of household characteristics and behavioural choices that may affect their overall welfare. Such analysis is expected to provide an important evidence base to be used when formulating and targeting better policies and strategies aimed at uplifting the standard of living of vulnerable households, particularly in developing countries with similar socio-economic conditions to Sri Lanka. In particular, the findings of this research provide valuable insights towards three Sustainable Development Goals (SDG) that are intended to be achieved by the year 2030 globally; no poverty, zero hunger and gender equality.

1.2 A Preview of the Book

This book consists of seven chapters, including this introductory chapter. Chapter 2 consists of two parts. The first part, which presents an overview of the consumption economies of scale, sets a conceptual foundation for the content of the book. The second part discusses the concept of equivalence scales, with special emphasis on the Engel equivalence scales method, as this is the main methodological approach employed in this book to measure economies of scale in food consumption.

Chapter 3 presents a detailed overview of Sri Lanka, followed by a detailed descriptive analysis of the data used in the study for empirical analysis. The data used in the empirical analysis of this book is sourced from the Household Income and Expenditure Survey (HIES) conducted by the Department of Census and Statistics (DCS) in Sri Lanka.

Chapter 4 seeks to address R.Q.1. The first part of this chapter presents a theoretical model on how and why food grown at home may generate economies of scale in consumption, followed by the empirical methodology adopted to estimate the magnitude of this effect. This chapter aims to provide empirical evidence on why low-income households, and rural households, in particular are able to achieve seemingly higher economies of scale in food consumption. The chapter also discusses the implications of home-grown food consumption in food security. The chapter concludes with a brief discussion of the policy implications arising from these findings.

Chapter 5 presents a theoretical model, supported by empirical findings, which is used to investigate R.Q.2 and R.Q.4 and explain the effects of domestic technology adoption. The empirical analysis in the chapter is further extended to examine the implications of household electricity access on economies of scale in food consumption, using the Engel equivalence scales, and thereby on poverty levels in Sri Lanka. The chapter ends with a brief discussion of policy implications of the empirical findings.

Chapter 6 seeks to answer R.Q.3 by investigating how the gender of the head of the household affects the household ability to reap the benefits of food consumption economies of scale. The chapter further extends the analysis to address R.Q.4 by examining the impact of these differences in food consumption economies of scale, based on the gender of the household headship, on the poverty status of households headed by males and females.

Chapter 7 consolidates the findings of the previous chapters and provides detailed discussion on policy implications arising from the findings of this book. This chapter also briefly presents the possible limitations of this study and points to future directions that could be explored to further investigate the phenomena that were at the heart of the study.

Chapter 2
Consumption Economies of Scale and Implications on Poverty Measurements

2.1 Introduction

In the previous chapter, the importance of incorporating economies of scale in consumption when measuring poverty was discussed. Furthermore, a number of knowledge gaps relating to the literature on consumption economies of scale were also identified. Before addressing this knowledge gap, it is important to understand the concept of consumption economies of scale and how these consumption economies of scale at the household level can be measured.

To this end, the objective of this chapter is threefold: (1) to discuss the concept and sources of consumption economies of scale; (2) to discuss how to measure consumption economies of scale and incorporate consumption economies of scale into the measurement of poverty using Engel equivalence scales; and (3) to discuss the implications of consumption economies of scale on poverty measurements. Section 2.2 discusses the concept of economies of scale in consumption and its sources. Section 2.3 presents a detailed discussion of the Engel equivalence scales method that can be used to measure consumption economies of scale and can also be used to adjust poverty measurements for potential consumption economies of scale. Section 2.4 explains the implications of consumption economies of scale on poverty measurements. Section 2.5 provides some concluding remarks.

2.2 Consumption Economies of Scale

As the old saying suggests, two people can live as cheaply as one; that is, the doubling of household size does not require a twofold increase in income to achieve the same level of standard of living (Deaton & Paxson, 1998; Nelson, 1988). Deaton (1997) documented that total household expenditure does not rise rapidly with increasing household size, leading to a decline in per capita expenditure (PCE) with increasing household size. This phenomenon is related to two factors: economies of scale in

M. Jayasinghe, *Poverty, Food Consumption, and Economic Development*,
https://doi.org/10.1007/978-981-16-8743-3_2

consumption and differences in the preferences/needs of the individuals in the household (Nelson, 1988). Economies of scale in consumption may emerge from joint consumption (sharing) of household public goods, bulk purchases, and increasing returns in household production of goods (Deaton & Paxson, 1998; Ermisch, 1981; Logan, 2011; Nelson, 1988). These sources of economies of scale are discussed in detail below.

2.2.1 Joint Consumption of Public Goods

The goods consumed in households belong to two broad categories: private and public. Private goods are those that provide satisfaction to individuals in the household. The consumption of a private good by one person in the household prevents the consumption of the same good by another member of the household. Often food and clothing have been cited as private goods in the literature (Kakwani & Son, 2005; Logan, 2011). In contrast, public goods are those that are jointly consumed within the household without compromising the consumption of another member in the household. Lazear and Michael (1980), for example, noted that "electric light in a room, the beauty of art work on the wall and the security provided by a locked bolt on the door" can be regarded as public goods. Joint consumption of these public goods by all household members prevents individuals in the household from purchasing their respective shares of the same good, significantly reducing their PCE to maintain the same level of standard of living.

When considering the empirical observations on economies of scale in consumption that arise from joint consumption, Bonke and Browning (2009) stated that married individuals enjoy higher lifetime gains from the publicness of some household expenditure. As such, larger households are assumed to enjoy identical material well-being at a lower PCE on public goods than that of smaller households. However, households can achieve economies of scale even for rivalrous goods up to a certain extent. Kakwani and Son (2005) for example, mention that clothing, although a private good, can be shared and passed on to some extent to the other household members. They further state that most of these household commodities are neither pure public nor pure private and that the degree of publicness or privateness varies with each commodity.

Many studies, attempting to measure the effect of joint consumption on economies of scale, have demonstrated that housing and household durable goods provide maximum consumption economies of scale in comparison to any other commodity (Deaton & Paxson, 1998; Kakwani & Son, 2005; Lazear & Michael, 1980; Logan, 2011; Nelson, 1988). In analysing United States consumer expenditure data, Nelson (1988) showed that the budget share on housing of a two-adult household is about 9% less than that of two single-adult households. Using Australian statistics, Kakwani and Son (2005) further confirmed that the expenditure share on housing decreases as household size increases, providing maximum economies of scale in consumption, compared to other household commodities. Logan (2011), in his study on testing the patterns of consumption economies of scale in households in the United States over time, revealed that the share of expenditure devoted to housing is negatively correlated with increases in household size.

Interestingly, many scholars have found that goods with higher private components also contribute to economies of scale up to a certain extent. The most extensively researched household private good in the literature on economies of scale is food. As noted previously, the consumption of food by one member in the household precludes the consumption of the same good by another household member. Therefore, the budget share on food is expected to increase concomitantly with the household size to maintain a given standard of living. This reflects the fact that the avenues for the economies of scale arising from joint consumption are rather limited for food. Nevertheless, many empirical studies have demonstrated the prevalence of economies of scale in food consumption, although it is less than those from public goods (Beatty, 2010; Becker, 1965; Byrne & Capps, 1996; Deaton & Paxson, 1998; Griffith et al., 2009; Jayasinghe et al., 2016, 2017, 2018; Lazear & Michael, 1980; Ma et al., 2006; Nelson, 1988; Vernon, 2005). For example, Deaton and Paxson (1998), using aggregate food expenditure data for the United States, United Kingdom, France, Taiwan, Thailand, Pakistan, and South Africa, showed that the aggregate per capita budget share on food and the household size demonstrate an inverse relationship. The study revealed that, at constant PCE, a unit increase in the logarithm of household size leads to a 5 percentage point decline in food budget share in poor countries.

Several studies have revealed the prevalence of economies of scale in other private goods such as clothing, entertainment, transportation, and health care. For example, Nelson (1988), using United States data, showed that private goods such as clothing and transportation contribute to economies of scale in consumption, although substantially less than those arising from housing and food. In analysing Australian data on clothing and footwear, alcohol and tobacco, education, personal care, and transport, Kakwani and Son (2005) showed that all commodities contribute to economies of scale in consumption in varing degrees, except for health care, where they assume that it is a pure private good. Logan (2011), further confirmed these observations.

2.2.2 *Bulk Purchases*

Bulk purchases and quantity discounts are important factors that lower the PCE of larger households (Nelson, 1988). Larger households may benefit from purchasing 'economy size' products and may enjoy the gains of promotional price discounts such as 'two-for-one'. Hence they pay less per unit or experience price reductions, particularly for perishable items. Deaton and Paxson (1998) documented that the economies of scale in food consumption through bulk purchases may partly explain the lower per capita food expenditure in larger households at constant PCE. Abdulai (2003), using Swiss data, confirmed Deaton and Paxson's (1998) observations, stating that economies of scale from buying in bulk, and thereby paying less per unit purchased in large households, probably explains the negative relationship between food share and the household size. In analysing British data, Griffith et al. (2009) revealed that bulk purchases save about 16% of the annual expenditure of an average household. They further stated that the amount saved varies with household size where larger households save more. Again, using British data, Beatty (2010) found that quantity discounts are economically important, because the larger households on average tend to pay less for food as they spend a significantly larger share of their expenditure on food items with bulk discounts or sales.

2.2.3 *Increasing Returns to Scale in Home Production*

The consumption economies of scale arising from increasing returns to scale in home production have been mainly explored using food expenditure sub-categories. The food expenditure sub-categories that have been explored in the literature include expenditure on food produced and consumed at home, and expenditure on food-away-from-home (Gan & Vernon, 2003; Jayasinghe et al., 2017; Logan, 2011; Ma et al., 2006; Vernon, 2005). These studies have revealed that each of these sub-categories of food expenditure contributes to economies of scale in rather different degrees. For example, using United States data, Magrabi et al. (1991) indicated that the household size elasticity for food consumed at home is less than 1, reflecting substantial economies of scale. Studies have consistently shown that the economies of scale in food consumed at home are derived mainly through increasing returns in household production. Gan and Vernon (2003) stated that larger households save time in preparing food at home. For example, while it may take 45 min to prepare a meal for one person, it may take just 1 h to prepare a meal for three. Accordingly, larger households are likely to have lower per capita time and monetary costs associated with dining at home, compared to smaller households (Gan & Vernon, 2003; McCracken & Brandt, 1987; Nelson, 1988; Stewart et al., 2004). Several studies have attempted to capture the economies of scale in consumption arising from increasing returns on household production using a 'household production framework' incorporating both the time and monetary cost of preparing food at home (Gan & Vernon, 2003; Van Hoa & Ironmonger, 1989; Vernon, 2005).

Vernon (2005), for example, using Russian data, found that economies of scale exist in food expenditure and food preparation time. The study indicated that a doubling of the household size leads to a decline in per capita food preparation time of about 75%. The results confirm the hypothesis that larger households substitute towards more 'time-intensive' food production technologies. More recently, Couprie and Ferrant (2012) stated that if an additional member adds relatively little to the time cost of household production, the per capita time spent on home production declines with household size. Using British data, the study found that two singles living separately need 6.3% more time to achieve the same utility level as individuals living as a couple and maintaining the same level of welfare. They estimated that a couple saves 2 h and 15 min daily by living together. However, Byrne and Capps

(1996), in analysing the food-away-from-home consumption of households in the United States showed that the strength of the negative relationship between household size and food consumed away-from-home has been declining in recent years. This is attributable to the fact that fast food industries have started providing more affordable 'family' offerings (or deals) for larger households.

2.3 Equivalence Scales

Household equivalence scales have been commonly used to quantify the relative income needed by households of different sizes and compositions to enjoy an identical welfare level. These equivalence scales (with strict assumptions applied in relation to the preferences and needs of the individuals in the households) also serve as an indicator of the degree of economies of scale in consumption.

An equivalence scale can be defined as "an index number [which]...indicates at reference prices the cost differential for a household, due to different household size, to reach the indifference curve of the reference household" (Grootaert, 1983). Deaton and Muellbauer (1980) presented an interesting comparison between equivalence scales and cost-of-living indices, stating that "equivalence scales are to welfare comparisons between households with different sizes, what cost-of-living indices are to welfare comparisons for a given household facing different prices". In practical contexts, OECD-modified equivalence scales are generally used by policymakers and statisticians, in particular to compare income inequality and poverty. The OECD-modified scales assign a value of 1 to the first household member, 0.5 to each additional adult, and 0.3 to each child.

There are two main problems associated with the use of these predefined equivalence scales such as the OECD-modified equivalence scales. Firstly, these scales do not take into account the differences in household socio-economic and demographic characteristics and assume that the economies of scale in consumption effect is the same for every additional child or an adult. For example, for each additional adult the scale is 0.5 and for each additional child it is 0.3, irrespective of how many additional adults, additional children, or household income, or location (e.g. rural versus urban). Secondly, no distinction is made for differences in economies of scale in consumption arising from various types of goods consumed in households. For example, not only may the economies of scale in food and non-food items be different, but differences may also be present between various food types (such as economies of scale in home-grown versus purchased food consumption). Therefore, it is necessary to consider a method of equivalence scales that is not complicated, yet enables adjustments in equivalence scales based on demographic, socio-economic and behavioural differences of households, where appropriate.

There are various econometric approaches to construct equivalence scales. These include the Engel (1857) method, the Rothbarth (1943) method, the Barten (1964) method, the use of unconditional equivalence scales as suggested in Pollak and Wales (1979), and the translation model based on Pollak and Wales (1981). The Rothbarth scales are based on a priori identification of goods that are consumed exclusively by adults and by children. An important assumption in the Rothbarth model is that the adult goods (i.e. the goods that are consumed by adults) are separable from those of children (Gronau, 1988, 1991). Therefore, this model appears to be inappropriate to identify the economies of scale in food consumption as 'food consumption' cannot be separated as an adult or non-adult good, and hence 'food' does not satisfy the main assumption of this model. The Barten model, also known as the public good approach, provides the theoretical basis for differences in the economies of scale in consumption of household public and private goods. Nevertheless, this method has not gained attention due to unavailability of the data, such as the price data, required to estimate equivalence scales. Other econometric methods also have not been commonly used in empirical studies due to methodological complications and data limitations. There is so far no consensus in the literature on the best method to estimate equivalence scales and all of these methods are based on different assumptions, require different data, and yield different sets of equivalence scales (Blackorby & Donaldson, 1991). The following section provides a critical discussion on the Engel method, the most popular method for the estimation of equivalence scales, and explains the rationale for adopting this method for the empirical analysis of this book.

2.3.1 Engel Method

Among these approaches, the Engel (1857) method, which is based on the idea that two households with identical food shares are equally well off, has been widely used in the literature (see, for example, Binh & Whiteford, 1990; Bosch-Domènech, 1991; Deaton & Muellbauer, 1986; Phipps & Garner, 1994; Seneca & Taussig, 1971; Tsakloglou, 1991; Watts, 1967). The Engel method, the first method to construct equivalence scales, dates back to the last century when Engel (1857) observed that the food expenditure share for richer households was lower than that of poorer households. He also observed that at the same level of income, the average propensity to consume food in smaller households is lower than that of larger households. Accordingly, Engel (1857) suggested that the budget share devoted to food could be used as an indicator of a household's standard of living and welfare. This suggests that two households with the same food share are expected to enjoy identical levels of welfare, irrespective of household size and income (Deaton & Muellbauer, 1980, 1986; Nicholson, 1976; Tsakloglou, 1991). Therefore, comparison of the income of those households at the same levels of food share would generate an index, known as equivalence scales, to explain the cost of maintaining a larger household relative to a smaller household. This cost reflects the cost of an additional household member.

Despite widespread application of the Engel method, it has been subject to criticisms by various scholars due to the lack of plausibility of its assumptions and the weaknesses of its theoretical base (Deaton & Muellbauer, 1980, 1986; Nicholson, 1976). Nicholson (1976) for example, argued that although food budget shares provide a numerical approach to compare the standard of living of households 'of the same composition', the indicator loses its intuitive appeal when comparing households of different compositions. In particular, the presence of a newborn child, largely a food-consuming individual, causes a higher budget share on food in that household compared to the time before the arrival of the child. Therefore, even if the household is fully compensated for the additional member, the higher food share of the household implies that their welfare has deteriorated. As a consequence, Deaton and Muellbauer (1986) asserted that under the Engel method, restoration of the food share of a household in the presence of an additional child implies 'overcompensation'. As such, they argue that the Engel scales overestimate the true cost of children. Additionally, the lack of an answer to Pollak and Wales' (1979) issue of under-identification has been identified as a problem of the Engel method in measuring economies of scale in consumption. Pollak and Wales (1979) argued that conditional equivalence scales that depend only on observed differences in consumption patterns of households with different compositions cannot be used for welfare comparisons. They further stated that unconditional equivalence scales that incorporate how parents feel about children should be used for welfare comparisons. Another major objection against the Engel method for measuring economies of scale in consumption relates to its implicit assumption of identical levels of economies of scale in consumption for every commodity. Hence, according to Deaton and Muellbauer (1986), this method lacks an adequate theoretical ground to estimate economies of scale in consumption.

Although the Engel method has its own drawbacks, the empirical analysis of this book is based on the Engel (1857) approach. This is not only because the model has received endorsement by Deaton (1981) as the most appropriate approach to estimate equivalence scales for developing countries using single cross sectional data, but also due to two main advantages associated with using this method to address the research questions that this book seeks to answer, as noted in Chap. 1.

Firstly, this book focuses on analysing economies of scale in food consumption. A major share of total expenditure is spent on food; hence it plays an integral part of consumption expenditure in poor households. The Engel method proves to be the best available approach to investigate the economies of scale in food consumption, as it is directly based on defining welfare through the expenditure share associated with food consumption. Secondly, the straightforwardness and simplicity in the estimation procedure of the Engel equivalence scales and its minimal data requirements increase the attractiveness of this method. More specifically, many developing countries, including Sri Lanka, do not have some of the required data, such as the price data needed for estimating the Barten model. In contrast, the estimation of Engel equivalence scales to measure the presence of economies of scale in food consumption requires only the expenditure on food and the total expenditure, which can be found in most household surveys.

2.3.2 Engel Method: Theoretical Foundation

As noted above, the Engel equivalence scales are based on the idea that two households with identical expenditure shares on food are equally well off. Following Deaton and Muellbauer (1980), the Engel approach can be explained as follows. For the purpose of convenience, assume that the reference household r consists of only one adult. Therefore, the equivalence scale can be considered as the number of 'adult equivalents' of the household under consideration.

The equivalence scale approach based on observed behaviour assumes that preferences are represented by a direct utility function as follows:

$$u = \upsilon(\boldsymbol{q}, z) \tag{2.1}$$

where u is the utility level of the household measured against the reference household, $\upsilon(.,.)$ is the common utility function, $\boldsymbol{q}$ is the vector of commodities purchased in the household under examination and z is the vector of demographic characteristics.

The cost function, with the minimum level of income X to reach a given level of utility u at price $\boldsymbol{p}$ and vector of demographic characteristics z, associated to utility function (2.1) is given by

$$c(u, \boldsymbol{p}, z) = X \tag{2.2}$$

This function then can be used as the basis for comparing welfare of households of different demographic compositions. If household r is considered as the reference household with $\boldsymbol{z^r}$demographic characteristics and utility level$u$, and with price vector$\boldsymbol{p}$, while household h is considered as the household under consideration with $\boldsymbol{z^h}$ demographic characteristics and utility level u, and with price vector $\boldsymbol{p}$, the ratio(μ^h) of the cost functions of the two households measures the relative cost of household h with $\boldsymbol{z^h}$ characteristics to achieve the same welfare (or utility) level as the reference household r with $\boldsymbol{z^r}$ characteristics. This ratio is known as the equivalence scale.

$$\mu^h = c(u, \boldsymbol{p}, \boldsymbol{z^h})/c(u, \boldsymbol{p}, \boldsymbol{z^r}) \tag{2.3}$$

Accordingly, Eq. (2.3) yields the equivalence scale corresponding to the utility function (2.1).

Equations (2.1), (2.2) and (2.3) are based on unobservable direct utility function. However, this can be related to observable behaviour through a cost function linking commodity expenditure to income, prices and demographic characteristics. Alternatively, following Deaton and Muellbauer (1980, pp. 40–43), in logarithmic form

$$w_i = \partial \ln c(u, \boldsymbol{p}, z)/\partial \ln p_i \tag{2.4}$$

where w_i is the budget share on good i. and p_i is the price of the ith commodity. The Eq. (2.4) yields the budget shares as a function of utility, prices and demographic variables. Since the utility levels are unobservable, the indirect utility function is derived to express utility in terms of income, prices, and demographic characteristics as follows:

$$u = \psi(X, \boldsymbol{p}, \boldsymbol{z}) \tag{2.5}$$

Substituting (2.5) into the right-hand side of Eq. (2.4) yields budget shares in terms of income, price and demographic characteristics, which are observable. It can be re-written as

$$w_i = f_i(X, p, z) \tag{2.6}$$

When formalising the Engel method, the cost function of any given household h with a vector $\boldsymbol{z}^h$ of demographic characteristics can be written as

$$c(u, \boldsymbol{p}, \boldsymbol{z}^h) = \mu(\boldsymbol{z}^h)c(u, \boldsymbol{p}) = X^h \tag{2.7}$$

where u is utility, $\boldsymbol{p}$ is prices, $c(u, \boldsymbol{p})$ is the cost function of household h, $\mu(z^h)$ is the number of equivalent adults in household h, and X^h is the income of household h. Consequently, the direct utility function is

$$u^h = u(q^h, z^h) = u[q^h/\mu(z^h)] \tag{2.8}$$

where u^h denotes utility and q^h denotes the vector of commodities consumed in the household h.

The per equivalent adult demand function takes the form of

$$q_i^h/\mu(z^h) = \phi_i[X^h/\mu(z^h), p] \tag{2.9}$$

where q_i^h is the quantity of the i th commodity consumed in the household h.

In budget share form, this becomes

$$w_i^h = \frac{p_i q_i^h}{X^h} = \frac{p_i \phi_i[X^h/\mu(\boldsymbol{z}^h), \boldsymbol{p}]}{[X^h/\mu(\boldsymbol{z}^h)]} \tag{2.10}$$

where w_i^h is the expenditure share on the ith commodity in household h. This is a function of both $X^h/\mu(\boldsymbol{z}^h)$ and not of X^h or $\mu(\boldsymbol{z}^h)$ separately. Accordingly, given that the price vector is the same for all the households, and both households (the household under consideration h and the reference household r) enjoy an identical foodshare w_f, according to the Engel supposition they are at the same level of welfare.

Hence Eq. (2.10) implies that

$$\frac{X^h}{\mu(z^h)} = \frac{X^r}{\mu(z^r)} \rightarrow \mu(z^h) = \frac{X^h}{X^r} \tag{2.11}$$

The equivalence scale for the reference household, with one adult, is $\mu(z^r)=1$ where r denotes the reference household, z^r is the vector of demographic characteristics of the reference household, and X^r is the income of reference household.

Based on Engel's view, in this basic model, the food share w_f is the indicator of welfare.[1] The food share is obtained by differentiating $\ln c(u, \boldsymbol{p}, z)$ with respect to $\ln p_f$(where p_f is the price of food):

$$w_f^h = \frac{\partial \ln c(u^h, \boldsymbol{p}^h, z^h)}{\partial \ln p_f} = \frac{\partial[\ln\mu(z^h) + \ln c(u^h, \boldsymbol{p}^h)]}{\partial \ln p_f} = \frac{\partial \ln c(u^h, \boldsymbol{p}^h)}{\partial \ln p_f} \tag{2.12}$$

where w_f^h is expenditure share on food in household h, p_f is the price of food, and u^h denotes the utility and $\boldsymbol{p}^h$ denotes vector of prices of household h.

Accordingly, with the assumption that prices are constant across all households, w_f^h changes with u^h and thus, it is considered as a proxy of welfare. The Engel method estimates the amount of money required to restore the food share to the previous level, where welfare is explained by the amount of food consumed in a household.

Consider for example, the total income of h (a household with couple and a child) and r (a household with a couple) in Sri Lanka are Sri Lankan Rupees (SLR) 30,000 and SLR 20,000 respectively. According to Eq. (2.11), the equivalence scale is simply 1.5 (calculated as 30,000/20,000). Accordingly, households with a couple and one-child require 50% more income than a childless couple to remain at the same welfare level. This can also be interpreted as the cost associated with one child relative to a childless couple.

Generally, the linear form of Engel curves is used to estimate Engel equivalence scales. This assumes that estimated equivalence scales are independent of household income. These are accurate only if preferences are homothetic and do not change over income dimensions. Nevertheless, household preferences appear to be non-homothetic in nature and several studies document clear evidence of nonlinearities in the Engel curves (Jayasinghe et al., 2016; Lewbel, 2008). This gives an indication that equivalence scales appear to depend on the chosen level of household income. Such a distinction, on whether equivalence scales, and thereby economies of scale in consumption, depend on household income or not, is an important consideration

[1] However, the same model can be extended to budget share for any commodity category. For example, Watts (1967) and Seneca and Taussig (1971) used an expenditure on a group of necessities: food, housing, clothing and transportation. Additionally, Binh and Whiteford (1990) used expenditure on food consumed at home, expenditure on all food, and expenditure on clothing, housing and fuel as their basket of necessities. Moreover, Phipps and Garner (1994) and Valenzuela (1996) used expenditure shares on food, clothing, housing and health care in their studies.

in the measurement of welfare. This is because, if it depends on household income, it implies that rich and poor households enjoy different levels of economies of scale even when their household size and composition are the same. A theoretical explanation of how and why household income may affect equivalence scales is presented below in Figs. 2.1 and 2.2.

Figure 2.1 depicts the cost of the household under examination, *h*, to enjoy the same level of welfare as the reference household, *r*. While in both households the food share declines with income, the larger household is required to spend more, namely $(X^h - X^r)$, to achieve the same level of welfare as for the smaller. The parallel linear Engel curves in Fig. 2.1 show that the distance between nominal incomes would be the same, irrespective of the welfare level chosen. Therefore, the equivalence scales based on linear Engel curves are independent of the household income at which

Fig. 2.1 Derivation of base-independent equivalence scales

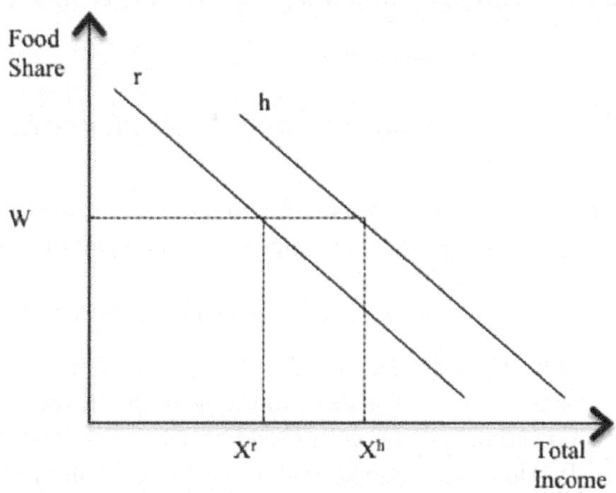

Fig. 2.2 Derivation of income-dependent equivalence scales

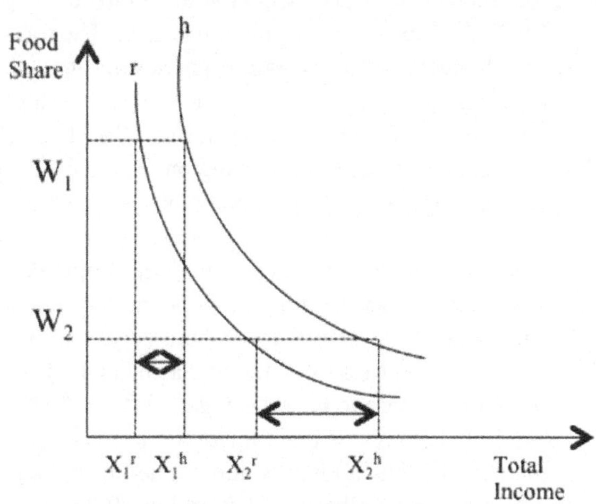

they are calculated. Such scales, known as "base-independent" equivalence scales, satisfy the property of "equivalent scale exactness" (Blackorby & Donaldson, 1991; Lewbel, 1989).

On the other hand, an important implication of non-linearity in Engel curves, as suggested by Lewbel (2008) and Jayasinghe et al. (2016), is that the cost of an additional household member (measured using equivalence scales) for the richer households can be different from that of the poorer households (e.g. Lewbel, 1989; Blackroby and Donaldson, 1991; Murthi, 1994; Donaldson & Pendakur, 2003). Due to non-linearity in Engel curves, the property of equivalence scale exactness may not hold as they may change with the (chosen) income level. Figure 2.2 graphically illustrates this phenomenon. As the distance between nominal income at welfare level W_1 and W_2 are different, the equivalence scale estimated at $W_1(X_1^h/X_1^r)$ differs from that of $W_2(X_2^h/X_2^r)$. In practice, income-dependent Engel equivalence scales are normally calculated using quadratic Engel curves.

2.3.3 *Estimation of Engel Equivalence Scales*

The first part of this sub-section discusses the estimation procedure of base-independent equivalence scales; the second part discusses that of income-dependent equivalence scales.

The estimation of Engel equivalence scales involves two steps.

Step 1: The first step involves estimation of the Engel curves. As this book focuses on economies of scale in food consumption, the Engel curve for food, taking the expenditure share on food and beverages as the dependent variable, is estimated. Two important points should be noted in relation to the Engel curve specifications for food employed in this study. Firstly, when estimating food Engel curves, following Deaton and Muellbauer (1986) a simple extension of the Working (1943) and Leser (1963) equation is used. The food share here is a linear function of the logarithm of per capita expenditure (PCE). Deaton and Muellbauer (1986) noted that this form frequently fits the data best, at least for the many household surveys from developing countries that they have examined. The specification of the Engel curve varies depending on whether the equivalence scales are assumed to be base-independent or whether they are allowed to vary with income. However, it is also worthwhile noting that, based on the Working (1943) and Leser (1963) model, the Engel's law means the logarithmic coefficient of PCE is less than 0. This functional form, although popular, does not account for the bounded nature of the food share, that is food share (w_f) can be greater than 0 and less than 1 ($0 < w_f < 1$). More specifically, a substantial increase in income may lead the food budget share to become negative (Gao, 2012). The estimation of non-linear food Engel curves, using a quadratic term of log of PCE may even increase the complexity of the above-mentioned bounded nature of the food share. Nevertheless, in this book, the use of Working (1943) and Leser (1963) functional form does not appear to raise such serious range violations, as the food budget shares used in this thesis range between 0 and 1 (i.e. $0 < w_f < 1$). Therefore, this functional form appears to be satisfactory as a local approximation. Secondly, in estimating food Engel curves, PCE is used as a proxy for per capita income (PCI) due to the enhanced reliability of expenditure data, as advocated by Deaton and Muellbauer (1980). As noted by Summers (1959), the Engel curve, when modelled on household expenditure, may suffer from endogeneity. To address this issue, the log of PCI was used as an instrumental variable (IV) for PCE. This is because income is highly correlated with household expenditure; however, as it is measured independently of expenditure it serves as a good IV for household expenditure (Deaton & Paxson, 1998; Lewbel, 1989).

Step 2: The second step involves substitution of the estimated coefficients into a mathematical equation, based on Engel's proposition that two households with the same food share are expected to enjoy identical levels of welfare irrespective of household size and/or income. Accordingly, a mathematical relationship is constructed such that $w^h = w^r$, where w^h is the budget share of a given household h^h and w^r is the budget share of the reference household h^r. While the substitution of the coefficients into the resulting mathematical equation is quite straightforward in the case of the linear Engel curve, it becomes more complex in the case of non-linear Engel curves.

The specifications of linear and non-linear Engel curves, as well as their estimation procedure, are discussed below.

2.3.3.1 Estimation of Base-Independent Engel Equivalence Scales

The specification of the linear Engel curve, with two demographic characteristics (adults and children), to estimate base-independent equivalence scales is as follows:

$$w_f = \beta_0 + \beta_1 \ln\frac{x}{n} + \gamma_1 n_a + \gamma_2 n_c + \varepsilon_i \tag{2.13}$$

where w_f refers to aggregate food budget share,n to household size (including children), n_a to number of adults and n_c to the number of children in the household. The variable x is the household expenditure, $\ln\frac{x}{n}$ is the logarithm of PCE and ε_i is the random error term. Accordingly, β_0, β_1, γ_1, and γ_2 are the parameters to be estimated. Following Deaton and Muellbauer (1986), a simple logarithmic transformation of PCE is used for two reasons. Firstly, the problem of heteroskedasticity commonly encountered with cross-sectional data can be reduced via such a transformation. Secondly, this form of transformation reduces the non-linearity in the data.

A two-adult and no-children household, (2,0) is used as the reference household, such that

$$w^r = \beta_0 + \beta_1 \ln\left(\frac{x^r}{n^r}\right) + \sum_{j=1}^{J} \gamma_j n_j^r \tag{2.14}$$

where $J = 2$ demographic groups: n_a is the number of adults, n_c is the number of children, and n^r is the household size in the reference household ($n^r = 2$).

and

$$w^h = \beta_0 + \beta_1 \ln\left(\frac{x^h}{n^h}\right) + \sum_{j=1}^{J} \gamma_j n_j^h \tag{2.15}$$

At equal expenditure shares, $w^h = w^r$, so

$$\beta_0 + \beta_1 \ln\left(\frac{x_r}{n_r}\right) + \sum_{j=1}^{J} \gamma_j n_j^{(r)} = \beta_0 + \beta_1 \ln\left(\frac{x_h}{n_h}\right) + \sum_{j=1}^{J} \gamma_j n_j^{(h)}$$

$$\beta_1 \left[\ln\left(\frac{x_r}{n_r}\right) - \ln\left(\frac{x_h}{n_h}\right)\right] = \sum_{j=1}^{J} \gamma_j n_j^{(h)} - \sum_{j=1}^{J} \gamma_j n_j^{(r)}$$

$$\ln\left(\frac{x_r}{n_r}\right) - \ln\left(\frac{x_h}{n_h}\right) = \sum_{j=1}^{J} \frac{\gamma_j}{\beta_1}\left(n_j^{(h)} - n_j^{(r)}\right)$$

$$\ln\left[\frac{x_r}{n_r} \Big/ \frac{n_h}{n_r}\right] = \sum_{j=1}^{j} \frac{\gamma_j}{\beta_1}\left(n_j^{(r)} - n_j^{(h)}\right)$$

$$\frac{x_h}{x_r} = \frac{n_h}{n_r} \exp \sum_{j=1}^{J} \frac{\gamma_j}{\beta_1}\left(n_j^{(r)} - n_j^{(h)}\right) \tag{2.16}$$

If the reference household is 2 adults with no children, as noted above, the equivalence scale using two demographic characteristics (i.e. n_a number of adults and n_c number of children) can be estimated using Eq. (2.17).

$$\frac{x_h}{x_r} = \frac{n_h}{2} \exp\left[\frac{\gamma_1}{\beta_1}(2 - n_a^{(h)}) - \frac{1}{\beta_1}\gamma_2 n_c^{(h)}\right] \tag{2.17}$$

where β_1, γ_1, and γ_2 are the estimated coefficients in the food Engel curve and $n_a^{(h)}$ and $n_c^{(h)}$ refer to the number of adults and children in the household under consideration, respectively. The equivalence scales calculated using this method are referred to as the base-independent equivalence scales.

2.3.3.2 Estimation of Income-Dependent Equivalence Scales

The specification of the Engel curve, with two demographic characteristics (adults and children), to estimate income-dependent equivalence scales, is as follows:

$$w_f = \beta_0 + \beta_1 \ln\frac{x}{n} + \beta_2(\ln\frac{x}{n})^2 + \gamma_1 n_a + \gamma_2 n_c + \gamma_3 n_a n_c \qquad (2.18)$$

where w_f refers to expenditure share on food, n to household size, n_a to number of adults and n_c to number of children in the household. The variable x is the household expenditure, $\ln\frac{x}{n}$ is the logarithm of PCE and $\left(\ln\frac{x}{n}\right)^2$ is the square of the logarithm of PCE. Here, in this case, another variable has been added to the model (i.e.$n_a n_c$), which is an interaction term of n_a andn_c, that estimates the joint effect of adults and children on food share. This is because in households which have both adults and children, their joint effect on the expenditure share on food may be different to the individual effect of adults and children on the expenditure share on food. Accordingly, $\beta_0, \beta_1, \beta_2, \gamma_1, \gamma_2$ and γ_3 are the parameters to be estimated.

The estimated coefficients of the quadratic Engel curves are used to construct income-dependent equivalence scales. Following Deaton (1981) and Lelli (2005), an analytical solution can be derived using a basic mathematical approach to solve quadratic equations, as follows. A two-adult and no-children household, (2, 0) is taken as the reference household, hence $n^r = 2$.

Consider the budget share on food in household h is

$$w_f^h = \beta_0 + \beta_1 \ln\left(\frac{x^h}{n^h}\right) + \beta_2(\ln\frac{x^h}{n^h})^2 + \gamma_1 n_a + \gamma_2 n_c + \gamma_3 n_a n_c \qquad (2.19)$$

and the budget share on food in the reference household r is

$$w_f^r = \beta_0 + \beta_1 \ln\left(\frac{x^r}{n^r}\right) + \beta_2(\ln\frac{x^r}{n^r})^2 + 2\gamma_1 \qquad (2.20)$$

Based on the Engel assumption that $w^h = w^r$, Eqs. (2.19) and (2.20) can be used to derive a quadratic equation as follows:

$$x^h = n^h \exp\left(\frac{-\beta_1 \pm \sqrt{\beta_1{}^2 - 4\beta_2 C}}{2\beta_2}\right) \qquad (2.21)$$

where $C = \gamma_1 n_a + \gamma_2 n_c + \gamma_3 n_a n_c - \beta_2\left(\ln\frac{x^r}{2}\right)^2 - \beta_1 \ln\left(\frac{x^r}{2}\right) - 2\gamma_1$

Equation (2.21) gives two roots for x^h for any given n^h, n_a^h and n_c^h. From the two possible solutions for x^h, the larger value of x^h corresponds to the relevant part (or downward sloping side) of the Engel curve. An equivalence scale is the ratio between

x^h and x^r, such that dividing x^h by the initially selected x^r yields the equivalence scale. Note that this ratio between x^h and x^r depends not only on n^h, n_a^h andn_c^h, but also on the income of the reference household, x^r. This is a natural result of the quadratic specification, with its implication that the costs imposed by household size are not proportionately the same at all levels of income. The empirical results and discussion in Chaps. 5 and 6 of this book are based on these income-dependent equivalence scales, estimated at the mean income.

2.4 Implications on Poverty Measurements

The approaches, process and data needed to measure poverty and gauge improvements in the prosperity of the poor people in each country present complex challenges. Poverty, in general, implies the lack of enough resources or abilities to meet the current needs of an individual. Often, the identification of the 'poor' involves a comparison of individuals' income or consumption with some predefined threshold below which individuals are considered as being poor in that particular attribute. Primarily, poverty used to be measured using monetary definitions, such as income and/or consumption expenditure. Although there is increasing recognition that poverty is multidimensional and that other aspects of people's well-being – such as health, education, and assets ownership—need to be considered too, this chapter focuses mainly on income and consumption. This is because the remaining chapters of this book focus on improving the measurement of monetary poverty.

2.4.1 *Choosing the Monetary Indicators of Poverty*

Income and consumption expenditure are the main indicators of monetary poverty measurement. The literature on monetary poverty argues that although the income data has its own merits in measuring poverty, consumption expenditure better suits as an indicator of poverty measurement than income for three reasons (see, for example, Christiaensen et al., 2002; Haughton and Khandker, 2009).

1. Consumption is a better outcome indicator than income as the actual consumption is a better indicator to gauge individual's ability to meet current basic needs. On the other hand, income indicates only the affordability of what is required to fulfil their basic needs but does not capture the other aspects, such as the accessibility and availability.
2. The income data collected from household surveys may be subject to recall errors, errors associated with income fluctuations, specifically informal sector incomes, errors associated with estimating income excluding the inputs purchased for agricultural production from the farmer's revenues, and underreporting of income.

3. Households in developing countries consume a substantial proportion of home-grown or freely received goods through informal exchange, which cannot be captured through income data.

Although the collection of expenditure data has its own challenges, it is widely accepted that consumption data may be more reliable and better reflects a household's actual standard of living and the ability to meet basic needs (Klasen, 1997). Rather than how much a particular household earned in a given period of time, consumption expenditure reflects how much of the income translated into actual consumption. This also captures household access to credits or savings to fulfil their consumption needs in the case of having insufficient income at a given time. The consumption data also takes the values of owner-occupied housing and home-grown and freely received goods (in the form of imputed values) into consideration in poverty measurements. For these reasons, the empirical analysis of this book is based on consumption expenditure.

2.4.2 Choosing Poverty Lines and Measurements

Once the monetary poverty indicator is identified at the household or individual level, the next consideration in the process of estimating poverty measures is to define the 'poverty line'. Poverty lines implies the cutoff points that separate the poor from the non-poor. When it comes to monetary poverty measurements, the cutoff points can be income or consumption. In some instances, multiple poverty lines are used to identify the presence of different levels of poverty.

There are two common poverty measures applied by researchers and policy analysts—relative poverty (based on the relative poverty line) and absolute poverty (based on the absolute poverty line).

Relative poverty: Relative poverty measures the level of poverty in relation to the overall distribution of income or consumption in a country; for example, the relative poverty line could be set at 50% of the country's mean income or consumption, and a household whose income or consumption falls below the specified threshold is considered as poor. Relative poverty enables the comparison of the lowest segment of a population with the upper segment (World Bank, 2013).

Absolute poverty: Absolute poverty measures the number of people or households living below a certain income/consumption threshold, which is known as the absolute poverty line. These poverty lines are often based on estimates of the cost of basic food needs to fulfil the minimum nutritional intake of household members with a provision for non-food needs also added. For example, if we consider one household and it is below the poverty level income (or expenditure), then that household is in absolute poverty. Considering that large parts of the populations of developing countries are struggling to fulfil their basic needs, absolute poverty measures are often considered as more suitable to measure poverty in developing countries than measuring relative poverty. As such, the empirical analysis of this book uses the absolute poverty line in Sri Lanka, known as the official poverty line (OPL). The foundation value of the OPL in Sri Lanka is SLR 1,423 and was determined by the DCS (2017) according to the 'cost of basic need method' on food and non-food consumption, based on the 2001/02 HIES series. It is based on 2030 kilocalories, the recommended daily per capita food energy requirement for Sri Lanka. The OPL in 2016 for Sri Lanka is SLR 4,166 (Department of Census & Statistics, 2017).

In addition to these two commonly used monetary poverty measures, 'subjective poverty' or 'self-reported measures of poverty' are also sometimes used (Goedhart et al., 1977; Pradhan & Ravallion, 2000).

Many alternative measures exist to measure poverty. The most widely used poverty index in the literature is the Foster Greer-Thorbecke Index (FGTI), which measures the incidence, depth and severity of poverty (Foster et al., 1984).

The FGTI is defined as:

$$P_\alpha = \frac{1}{n}\sum_{i=1}^{q}\left(\frac{z - y_i}{z}\right)^\alpha$$

where n is total population, $z$is poverty line, y_i is income or consumption expenditure of the ith individual, q is the number of persons with income or expenditure below the poverty line, and α is a measure of sensitivity of the index to poverty.

If $\alpha = 0$, the FGTI reduces to the Headcount Index (HI), if $\alpha = 1$, the index provides the Poverty Gap Index (PGI) and if $\alpha = 2$, the index provides the Squared Poverty Gap Index (SPGI) (Foster et al., 1984).

$$HI = \frac{1}{n}\sum_{i=1}^{q}(1) = \frac{q}{n}$$

$$PGI = \frac{1}{n}\sum_{i=1}^{q}\left(\frac{z - y_i}{z}\right)$$

$$SPGI = \frac{1}{n}\sum_{i=1}^{q}\left(\frac{z - y_i}{z}\right)^2$$

The HI indicates the incidence of poverty. This is the share of the population whose income or consumption is below the poverty line or the share of the population that cannot afford to buy a basic basket of goods. PGI indicates the depth of poverty. This provides information regarding how far off households are from the poverty line. SPGI shows the severity of poverty and takes into account not only the distance separating the poor from the poverty line (the poverty gap), but also the inequality among the poor. The empirical analysis of this book mainly uses the HI measurement.

Table 2.1 Identification of 'poor' households

PCE method			EPEA method		
Household size	Per capita household expenditure	Poverty status	Household equivalence scale (OECD modified)	Expenditure per adult equivalent	Poverty status
(1)	(2)	(3)	(4)	(5)	(6)
2 adults and 1 child = 3	12,000/3 = 4000	Poor	2 adults 1 child = 1 + 0.5 + 0.3 = 1.8	12,000/1.8 = 6,667	Non-poor

2.4.3 Implications of Economies of Scale

Irrespective of the indicator of monetary poverty (income or consumption) used, it is typically necessary to aggregate information provided at the household or individual level for many sources of income or consumption in the survey. This aggregation is a complex task which requires various adjustments. One crucial adjustment required, as discussed in the previous sections of this chapter is to adjustments to reflect the economies of scale in consumption, so that the poverty of large households may not be overestimated.

Table 2.1 explains the implications of consumption economies of scale on poverty measurements using a hypothetical example of a household of 2 adults and 1 child with expenditure per month totalling SLR 12,000. In general, under the conventional methods the poverty figures are estimated using the PCE. For example, the percentage of poor households is estimated using the number of households whose PCE falls below the official poverty line. In this example, the size of a household of 2 adults and 1 child is 3 and the per capita household expenditure is SLR 4,000 can be obtained by dividing the total expenditure by household size. Based on the official poverty line in Sri Lanka in 2016, SLR 4,166 real total expenditure per person per month, this particular household is considered as 'poor'.

However, if necessary allowances for consumption economies of scale are incorporated into poverty measurements, household expenditure should be divided by the household specific equivalence scale to obtain expenditure per equivalent adult (EPEA) (instead of the PCE), which will then be compared against the official poverty line. For simplicity of illustration, here OECD-modified equivalence scales are used. The OECD-modified scales assign a value of 1 to the first household member, 0.5 to each additional adult, and 0.3 to each child (Eurostat, 2014). Accordingly, as can be seen in Column 4, the household equivalence scale is 1.8. In this example, the EPEA is SLR 6,667 and hence, the same household that was identified as 'poor' based on the PCE will be considered as 'non-poor' when economies of scale in consumption are incorporated. This explains how the conventional poverty figures based on per capita methods may overestimate the poverty level of large households.

2.5 Concluding Remarks

The objective of this chapter was to discuss the concept and the three sources of consumption economies of scale; sharing of household goods, the bulk purchases, and the increasing returns to scale in home production. Usually, household equivalence scales are used to quantify the extent of consumption economies of scale. This chapter discussed the most popular method to estimate equivalence scale, the Engel method. The empirical analysis in Chaps. 5 and 6 of this book also uses the Engel method due to two reasons; Firstly, the Engel method appears to be the best approach to investigate the economies of scale in food consumption as it defines household welfare in terms of the expenditure share on food. Secondly, the simplicity in the estimation procedure and the minimalistic data requirements enable exploration of some important aspects relating to economies of scale in food consumption that have not been found in the literature to date. The chapter also discusses the implications of economies of scale in consumption on poverty measurement using a hypothetical example. This aspect will be further discussed in Chaps. 5 and 6 of this book.

This book uses Sri Lanka as a case study. The next chapter discusses the demographic and socio-economic characteristics of households in Sri Lanka and presents a detailed description of the data used in the subsequent empirical analysis.

References

Abdulai, A. (2003). Economies of scale and the demand for food in Switzerland: Parametric and non-parametric analysis. *Journal of Agricultural Economics, 54*(2), 247–267. https://doi.org/10.1111/j.1477-9552.2003.tb00062.x

Barten, A. P. (1964). Family composition, prices and expenditure patterns. In P. Hart & G. Mills (Eds.), *Economic Analysis for National Economic Planning*. London: Butterworth.

Beatty, T. K. M. (2010). Do the poor pay more for food? evidence from the United Kingdom. *American Journal of Agricultural Economics, 92*(3), 608–621.

Becker, G. S. (1965). A theory of the allocation of time. *The Economic Journal, 75*(299), 493–517.

Binh, T. N., & Whiteford, P. (1990). Household equivalence scales—new Australian estimates from the 1984 Household Expenditure Survey. *Economic Record, 66*(194), 221–234.

Blackorby, C., & Donaldson, D. (1991). Adult equivalence scales, interpersonal comparisons of well-being, and applied welfare economics. In J. Elster & J. E. Roemer (Eds.), *Interpersonal Comparisons of Well-Being* (pp. 164–199). Cambridge University Press.

Bonke, J., & Browning, M. (2009). The distribution of financial well-being and income within the household. *Review of Economics of the Household, 7*(1), 31–42.

Bosch-Domènech, A. (1991). Economies of scale, location, age, and sex discrimination in household demand. *European Economic Review, 35*(8), 1589–1595. https://doi.org/10.1016/0014-2921(91)90020-J

Byrne, P. J., & Capps, O. (1996). Does Engel's law extend to food away from home. *Journal of Food Distribution Research, 27*, 22–32.

Christiaensen, L., Scott, C., and Wodon, Q. (2002). Poverty measurement and analysis. University Library of Munich, Germany.

Couprie, H., and Ferrant, G. (2012). Welfare comparisons, economies of scale and equivalence scale in time use. *Thema Working Paper*. France: Université de Cergy Pontoise.

Deaton, A. (1981). Three essays on Sri Lankan household survey. *Living Standard Measurement Study Working Paper no.11*. Washington DC: World Bank.
Deaton, A. (1997). *The analysis of household surveys*. The Johns Hopkins University Press.
Deaton, A., & Muellbauer, J. (1980). *Economics and consumer behavior*. Cambridge University Press.
Deaton, A., & Muellbauer, J. (1986). On measuring child costs: With applications to poor countries. *The Journal of Political Economy, 94*(4), 720–744. https://doi.org/10.1086/261405
Deaton, A., & Paxson, C. (1998). Economies of scale, household size, and the demand for food. *The Journal of Political Economy, 106*(5), 897–930.
Department of Census and Statistics. (2017). Poverty indicators. Colombo: Department of Census and Statistics.
Donaldson, D., & Pendakur, K. (2003). Equivalent-expenditure functions and expenditure-dependent equivalence scales. *Journal of Public Economics, 88*, 175–208.
Engel, E. (1857). Die lebenskosten belgischer arbeiter-familien fruher und jetzt. *International Statistical Institute Bulletin, 9*, 1–74.
Ermisch, J. F. (1981). An economic theory of household formation: Theory and evidence from the general household survey. *Scottish Journal of Political Economy, 28*(1), 1–19.
Eurostat. (2014). Glossary: equivalised disposable income. Retrieved August 01, 2016, from http://ec.europa.eu/eurostat/statistics-explained/index.php/Glossary:Equivalised_disposable_income
Foster, J. E., Greer, J., & Thorbecke, E. (1984). Notes and comments a class of decomposable poverty measures. *Econometrica, 52*(3), 761–766.
Gan, L., & Vernon, V. (2003). Testing the Barten model of economies of scale in household consumption: Toward resolving a paradox of Deaton and Paxson. *The Journal of Political Economy, 111*(6), 1361–1377. https://doi.org/10.1086/378534
Gao, G. (2012). World food demand. *American Journal of Agricultural Economics, 94*(1), 25–51.
Goedhart, T. V., Harberstadt, A. K., & van Praag, B. M. S. (1977). The poverty line: Concept and measurement. *Journal of Human Resources, 12*(4), 503–520.
Griffith, R., Leibtag, E., Leicester, A., & Nevo, A. (2009). Consumer shopping behavior: How much do consumers save? *The Journal of Economic Perspectives, 23*(2), 99–120.
Gronau, R. (1988). Consumption technology and the intrafamily distribution of resources: Adult equivalence scales reexamined. *The Journal of Political Economy, 96*(6), 1183–1205. https://doi.org/10.1086/261583
Gronau, R. (1991). The intrafamily allocation of goods-how to separate the adult from the child. *Journal of Labor Economics, 9*(3), 207–235.
Grootaert, C. (1983). The conceptual basis of measures of household welfare and their implied survey data requirements. *The Review of Income and Wealth, 29*(1), 1–21.
Haughton, J., & Khandker, S. R. (2009). *Handbook on poverty and inequality*. World Bank.
Jayasinghe, M., Chai, A., Ratnasiri, S., & Smith, C. (2017). The power of the vegetable patch: How home-grown food helps large rural households achieve economies of scale and escape poverty'. *Food Policy, 73*, 62–74.
Jayasinghe, M., Ratnasiri, S., Smith, C., & Chai, A. (2018). Domestic technology, consumption economies of scale and poverty: Evidence from Sri Lanka. *Applied Economics, 50*, 1777–1789.
Jayasinghe, M., Smith, C., Chai, A., & Ratnasiri, S. (2016). The implications of income dependent equivalence scales for measuring poverty in Sri Lanka. *International Journal of Social Economics, 43*, 1300–1314.
Kakwani, N., & Son, H. H. (2005). Economies of scale in household consumption: With application to Australia. *Australian Economic Papers, 44*(2), 134–148. https://doi.org/10.1111/j.1467-8454.2005.00254.x
Klasen, S. (1997). Poverty, inequality and deprivation in South Africa: An analysis of the 1993 Saldru Survey. *Social Indicators Research, 41*(1/3), 51–94.
Lazear, E. P., & Michael, R. T. (1980). Family size and the distribution of real per capita income. *The American Economic Review, 70*(1), 91–107.

Lelli, S. (2005). Using functionings to estimate equivalence scales. *Review of Income and Wealth, 51*(2), 255–283.
Leser, C. E. V. (1963). Forms of Engel functions. *Econometrica, 31*(4), 694–703.
Lewbel, A. (1989). Household equivalence scales and welfare comparisons. *Journal of Public Economics, 39*(3), 377–391. https://doi.org/10.1016/0047-2727(89)90035-2
Lewbel, A. (2008). Engel curves. In S. N. Durlauf & L. E. Blume (Eds.), *The New Palgrave Dictionary of Economics* (2 ed.). Basingstoke: Palgrave Macmillan.
Logan, T. D. (2011). Economies of scale in the household: Puzzles and patterns from the American past. *Economic Inquiry, 49*(4), 1008–1028.
Ma, H., Huang, J., Fuller, F., & Rozelle, S. (2006). Getting rich and eating out: Consumption of food away from home in urban China. *Canadian Journal of Agricultural Economics, 54*(1), 101–101.
Magrabi, F. M., Chung, Y. S., Cha, S. S., & Yang, S.-J. (1991). *The economics of household consumption*. Greenwood Press.
McCracken, V. A., & Brandt, J. A. (1987). Household consumption of food-away-from-home: Total expenditure and by type of food facility. *American Journal of Agricultural Economics, 69*(2), 274–284.
Murthi, M. (1994). Engel equivalence scales in Sri Lanka: exactness, specification, measurement error. In R. Blundell, I. Preston & I. Walker (Eds.), *The Measurement of Household Welfare*. Cambridge: Cambridge University Press.
Nelson, J. A. (1988). Household economies of scale in consumption: Theory and evidence. *Econometrica, 56*(6), 1301–1314.
Nicholson, J. L. (1976). Appraisal of different methods of estimating equivalence scales and their results. *Review of Income and Wealth, 22*(1), 1–11. https://doi.org/10.1111/j.1475-4991.1976.tb01138.x
Phipps, S. A., & Garner, T. (1994). Are equivalence scales the same for the United States and Canada? *Review of Income and Wealth, 40*(1), 1–17. https://doi.org/10.1111/j.1475-4991.1994.tb00042.x
Pollak, R. A., & Wales, T. J. (1979). Welfare comparisons and equivalence Scales. *The American Economic Review, 69*(2), 216–221.
Pollak, R. A., & Wales, T. J. (1981). Demographic variables in demand analysis. *Econometrica, 49*(6), 1533–1551.
Pradhan, M., & Ravallion, M. (2000). Measuring poverty using qualitative perceptions of welfare. *Review of Economics and Statistics, 82*(3), 62–71.
Rothbarth, E. (1943). Note on a method of determining equivalent income for families of different composition. In C. Madge (Ed.), *War-time Pattern of Saving and Spending*. Cambridge: Cambridge University Press.
Seneca, J. J., & Taussig, M. K. (1971). Family equivalence scales and the personal income tax exemptions for childen. *The Review of Economics and Statistics, 53*(3), 253–262.
Stewart, H., Blisard, N., Bhuyan, S., and Nayga, R. M. (2004). The demand for food-away-from-home: full-service or fast food? *Agricultural Economic Report*. Washington DC: United States Department of Agriculture.
Summers, R. (1959). A note on least squares bias in household expenditure analysis. *Econometrica, 27*, 121–126.
Tsakloglou, P. (1991). Estimation and comparison of two simple models of equivalence scales for the cost of children. *The Economic Journal, 101*(405), 343–357.
Valenzuela, M. R. J. (1996). Engel scales for Australia, the Philippines and Thailand: A comparative analysis. *The Australian Economic Review, 114*(2), 189–198. https://doi.org/10.1111/j.1467-8462.1996.tb00924.x
Van Hoa, T., & Ironmonger, D. S. (1989). Equivalence scales: A household production approach. *Economics Letters, 31*(4), 407–410.
Vernon, V. (2005). *Food expenditure, food preparation time and household economies of scale*. State University of New York.

Watts, H. (1967). The ISO-prop index: An approach to the determination of differential poverty income thresholds. *Journal of Human Resources, 2*(1), 3–18.

Working, H. (1943). Statistical laws of family expenditure. *Journal of the American Statistical Association, 38*(221), 43–56.

World Bank. (2013). Poverty. Retrieved March 05, 2021, from http://www.worldbank.org/en/topic/poverty

Chapter 3
Socio-economic and Demographic Characteristics of Sri Lankan Households

3.1 Introduction

The empirical analysis undertaken in this book utilises data from the Household Income and Expenditure Survey (HIES) for Sri Lanka. This chapter presents a detailed description of the data with a view of providing an understanding about socio-economic context in Sri Lanka.

This chapter is organised as follows. Section 3.2 provides an overview of Sri Lanka. Section 3.3 presents a detailed description of the five series of HIES data, with particular focus on demographic characteristics, household expenditure, and the status of poverty in the country. Section 3.4 provides concluding remarks.

3.2 An Overview of Sri Lanka

Sri Lanka (formerly known as Ceylon) is an island located at the southern tip of the Indian subcontinent. The Island covers a land area of 65,610 square km, 1,340 km of which is coastline. After gaining independence from British colonial rule in 1948, Sri Lanka has experienced an intriguing mix of successes and failures in the social, political and economic contexts. Despite its low income level, Sri Lanka boasted high social and human welfare indicators such as literacy rate (92.9%), life expectancy (75.5 years), and an infant mortality rate of 8.5 per 1,000 live births in 2020 comparable to economies with much higher incomes (Central Bank of Sri Lanka, 2020). The human development index value reported for Sri Lanka is 0.782, ranking it 72nd amongst 189 countries in the world, and demonstrating that Sri Lanka's performance in broader welfare indicators is well above its South Asian neighbours (United Nations Development Program (UNDP), 2020). The country suffered from 26 years of bloody civil conflict between the Liberation Tigers of Tamil Eelam (LTTE) and the Sri Lankan government, interrupting sustained socio-economic development in the

M. Jayasinghe, *Poverty, Food Consumption, and Economic Development*,
https://doi.org/10.1007/978-981-16-8743-3_3

country. Even after a decade since the end of the war, the country is still experiencing significant development challenges.

As can be seen in Fig. 3.1, Sri Lanka consists of 9 administrative provinces, which are further subdivided into 25 districts. Table 3.1 presents the distribution of population by province and district in 2001 and 2012.

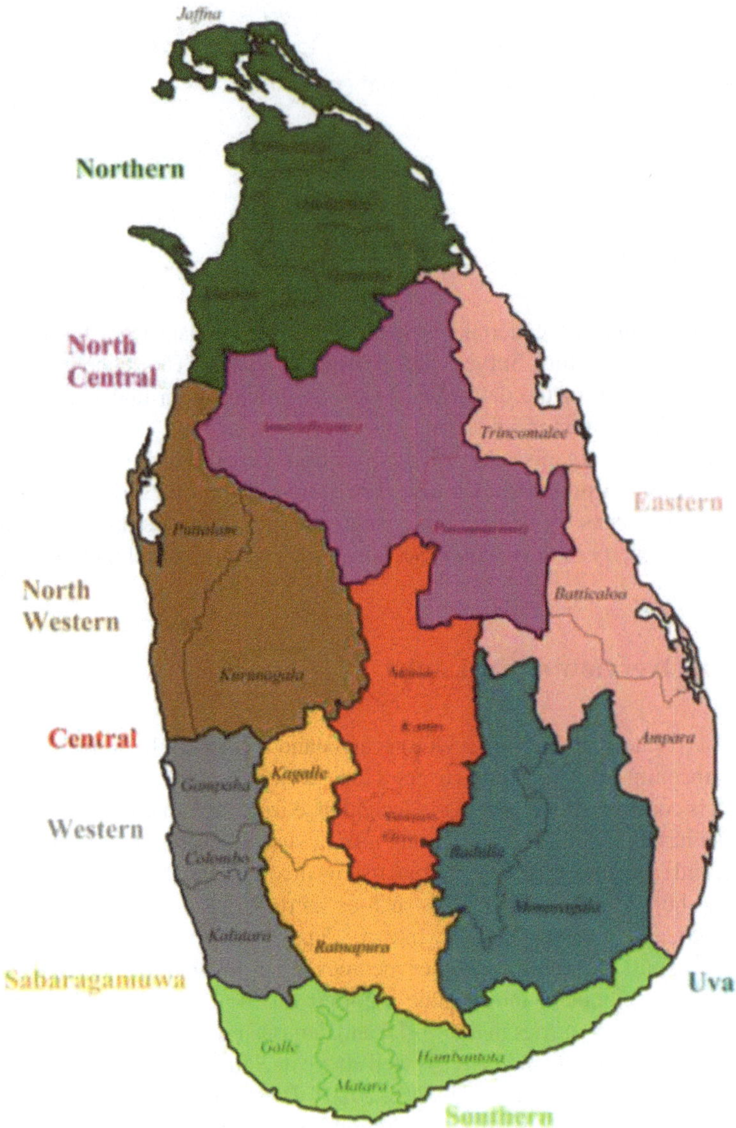

Fig. 3.1 Map of Sri Lanka, by district and province. *Source* Department of Census and Statistics (2011)

Table 3.1 Population, by province and district

Province	District	Percentage of total population (%)	
		2001	2012
(1)	(2)	(3)	(4)
Western		28.6	28.7
	Colombo	12.0	11.4
	Gampaha	11.0	11.3
	Kalutara	5.7	6.0
Central		12.9	12.6
	Kandy	6.8	6.8
	Matale	2.3	2.4
	Nuwara Eliya	3.7	3.5
Southern		12.1	12.2
	Galle	5.3	5.2
	Matara	4.1	4.0
	Hambantota	2.8	2.9
Northern		5.5	5.2
	Jaffna	2.6	2.9
	Mannar	0.8	0.5
	Vavuniya	0.8	0.8
	Mullaitivu	0.6	0.5
	Kilinochchi	0.7	0.6
Eastern		7.6	7.7
	Batticaloa	2.6	2.6
	Ampara	3.2	3.2
	Trincomalee	1.8	1.9
North-Western		11.5	11.7
	Kurunegala	7.8	7.9
	Puttalam	3.8	3.7
North Central		5.9	6.2
	Anuradhapura	4.0	4.2
	Polonnaruwa	1.9	2.0
Uva		6.3	6.2
	Badulla	4.1	4.0
	Monaragala	2.1	2.2
Sabaragamuwa		9.6	9.5
	Ratnapura	5.4	5.3
	Kagalle	4.2	4.1

(continued)

Table 3.1 (continued)

Province	District	Percentage of total population (%)	
		2001	2012
(1)	(2)	(3)	(4)
Sri Lanka		100	100

Note A number of Districts in the Northern and Eastern provinces were excluded during the 2001 census due to the prevailing civil conflict during the period. The population of these districts is an estimate

Source Department of Census and Statistics (2012)

Sri Lanka is divided into three sub-national regions, generally known as sectors: Urban, Rural and Estate. The DCS (2011) defines the areas governed by either municipal council or urban council as the urban sector. The estate sector comprises tea plantation areas which are more than 20 acres in extent and which have a minimum of 10 residential labourers. The residential areas that do not belong to either the urban sector or the estate sector, are classified as the rural sector. Accordingly, each district, listed in Table 3.1 may include urban, rural, and estate sectors. Table 3.2 presents the percentage of population in each district by sector.

3.3 Data—Household Income and Expenditure Survey (HIES)

This section presents a detailed analysis of household demographic and socio-economic characteristics and their intertemporal trends, using five series of HIES, conducted by the DCS in Sri Lanka. HIES is usually conducted every 5 years, with the first HIES conducted in 1990/1991. The survey uses a two-stage stratified sample design and the district is the main domain used for the stratification. The urban, rural and estate sectors in each district of the country are the second selection domains. The data was collected from an Island-wide representative sample during twelve consecutive monthly rounds, in order to capture seasonal variations in income, expenditure and consumption of households. Table 3.3 summarises the sample size and coverage of each survey.

Both the 1990/1991 and 2002 surveys covered only 17 districts out of 25 districts, excluding the Northern (which covers Jaffna, Mannar, Vavuniya, Mullativu, and Killinochchi districts) and Eastern (which covers Batticaloa, Ampara, and Traincomalee) provinces, due to the prevailing civil war condition in the area during that period. This excludes about 13% of the total population in the country from the sample. The 2009/2010 survey, however, covers 22 districts, excluding only the Mannar, Mullathivu, and Kilinochchi districts in the Northern province, due to landmine clearance and resettlement activities that took place after the end of the civil

Table 3.2 Population by district and sector—2012

District	Percentage of population (%)		
	Urban	Rural	Estate
(1)	(2)	(3)	(4)
Colombo	77.6	22.1	0.3
Gampaha	15.6	84.3	0.1
Kalutara	8.9	88.0	3.1
Kandy	12.4	81.4	6.2
Matale	12.4	83.6	3.9
Nuwara Eliya	5.6	40.9	53.5
Galle	12.5	85.7	1.8
Matara	11.9	85.4	2.8
Hambantota	5.3	94.7	–
Jaffna	20.1	79.9	–
Mannar	24.5	75.5	–
Vavuniya	20.2	79.8	–
Mullaitivu	–	100.0	–
Kilinochchi	–	100.0	–
Batticaloa	28.7	71.3	–
Ampara	23.6	76.4	–
Trincomalee	22.4	77.6	–
Kurunegala	1.9	97.7	0.5
Puttalam	8.8	91.0	0.2
Anuradhapura	5.9	94.1	–
Polonnaruwa	–	100.0	–
Badulla	8.6	72.6	18.9
Monaragala	–	98.1	1.9
Ratnapura	9.1	81.7	9.2
Kagalle	1.9	91.3	6.8
Sri Lanka	18.2	77.4	4.4

Source Department of Census and Statistics (2012)

conflict. Only about 1.6% of the total population in the country lives in these districts. Both 2012/2013 and 2016 surveys cover all 25 districts in the country.

The survey data was collected through direct interview using a survey questionnaire. The questionnaire covers (a) individual data on demographic characteristics of all household members, (b) household expenditure (on food and non-food), and (c) income received by each household member in the form of monetary and non-monetary incomes.

Table 3.3 Summary of survey samples

Survey series	Survey period	Number of households	Number of districts surveyed
HIES 1990/1991	June 1990–May 1991	18,462	17
HIES 2002	Jan 2002–Dec 2002	16,924	17
HIES 2009/2010	July 2009–June 2010	19,958	22
HIES 2012/2013	June 2012–May 2013	20,540	25
HIES 2016	Jan 2016–Dec 2016	21,756	25

Source Author's compilation based on Department of Census and Statistics (1993, 2003, 2011, 2014, 2017)

3.3.1 Demographic Characteristics

The household is the unit of analysis in all of the chapters in this book. A household can be either a one-person household or a multi-person household. A one-person household is a unit where a person lives by him/herself and makes separate provision for his/her food, whereas a multi-person household is a group of two or more persons who live together and have common meal arrangements. Furthermore, boarders and servants who share the meals and housing facilities with other members of the household are also considered as members of the household (Department of Census and Statistics, 2011). In the household, the head of the household is usually expected to make household decisions. The head of the household is defined as a person who resides in the household and who is accepted by the other members of the household as the head (Department of Census and Statistics, 2011). Table 3.4 presents a description of the basic demographic and social characteristics of the head of the household, by sector and by survey year.

As shown in Table 3.4, the average household size has been gradually declining in Sri Lanka. For example, the average household size has gradually declined from 4.9 in 1990/1991 to 3.8 in 2016. Household headship by gender is an important consideration in this book, in particular because the gender of the head of the household has gained significant attention in the literature on poverty and welfare. Firstly, there is a high incidence of female-headed households in Sri Lanka (Jayasinghe and Smith 2021). De Silva (2005) noted that Sri Lanka has the highest proportion of female-headed households in South Asia. The political unrest in the south of the country in the early 1980s, and the almost 30 year long civil war originating in the north

Table 3.4 Socio-demographic characteristics of the head of the household, by survey year

Characteristics of household head	1990/1991	2002	2009/2010	2012/2013	2016
(1)	(2)	(3)	(4)	(5)	(6)
Average household size	4.9	4.2	4	3.9	3.8
Female-headed households	20.1	20.7	24.1	23.5	25.8
Age group (%)					
18–39	27.5	26.1	23.4	22.7	20
40–59	46	50.5	49.3	48.7	47
60 and above	26.5	23.4	27.3	28.6	33
Level of education (%)					
No schooling	11.4	6.7	5.6	3.4	3.4
Up to grade 5	40.9	33.4	30.8	23.2	21.8
Passed grade 6–10	31.1	36.8	39.6	46.4	47.1
Passed G.C.E. O/L	11.4	11.7	9.9	15.3	14.8
Passed G.C.E A/L	3.8	9.3	11.8	9	9.9
Passed degree & above	1.4	2.1	2.4	2.5	2.9
Marital status (%)					
Never married	2.7	2.4	2.2	2.1	2.3
Married	79.4	81.3	78.4	79	77.5
Widowed	16.5	14.3	17.0	15.9	16.7
Divorced/separated	1.4	2	2.4	3	3.4

Source Author's compilation based on Department of Census and Statistics (1993, 2003, 2011, 2014, 2017)

and east have resulted in a significant number of households emerging as female-headed households due to widowhood. The international out-migration of the male head for economic reasons has also partly contributed to an increased proportion of female-headed households in Sri Lanka (Ruwanpura and Humphries, 2004). As can be seen in Table 3.4, during the periods of 1990/1991 and 2001/2002, about one fifth of the households in the sample are headed by females and since 2009/2010 about one fourth of households heads are females.

The empirical analysis of this book is completed using the data on household expenditure, particularly the expenditure on food. The next section provides an in-depth explanation of household expenditure patterns of Sri Lankan households.

3.3.2 *Household Expenditure*

Household expenditure data consists of expenditure on food and beverages, alcohol and tobacco, and expenditure on non-food items. All of these expenditure data are

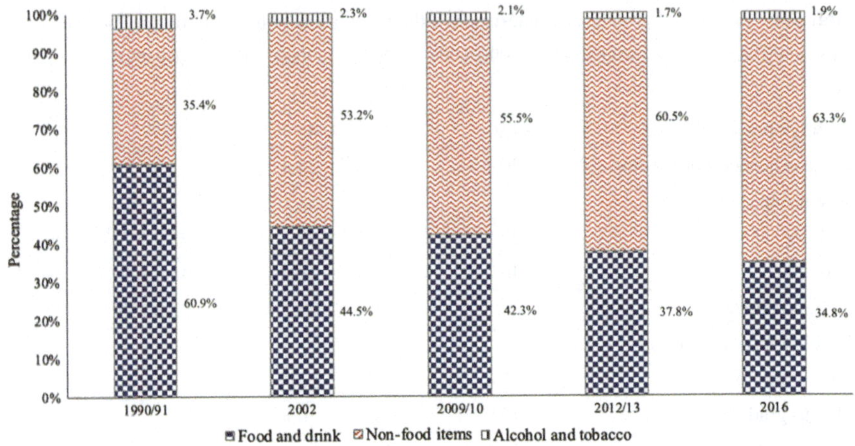

Fig. 3.2 Household expenditure share on food, non-food and alcohol, by survey year. *Source* Author's compilation based on Department of Census and Statistics (1993, 2003, 2011, 2014, 2017)

collected and recorded at the household level. Household expenditure has commonly been identified as a proxy for household income due to the accuracy of measurements (Deaton & Muellbauer, 1980).

Figure 3.2 presents the household average expenditure share on food, non-food items, and alcohol and tobacco, as a percentage of total household expenditure. As can be seen in Fig. 3.2, the expenditure share on food has declined over time from 60.9% in 1990/1991 to 34.8% in 2016. On the other hand, the expenditure share on non-food items has increased over time from 35.4% in 1990/1991 to 63.3% in 2016. In 2016, about 2% of household budget was allocated on alcohol and tobacco compared to 3.7% in 1990/1991.

Food expenditure

Food expenditure consists of expenditure on food and nonalcoholic beverages consumed by the household. It covers a total of 249 commodity items, classified under the 18 sub-categories: 'cereal', 'prepared food', 'pulses', 'vegetables', 'yams', 'meat', 'fish', 'dried fish', 'eggs', 'coconuts', 'condiments', 'other foods', 'milk and milk products', 'fats and oils', 'sugar', 'fruits', 'confectionery' and 'non-alcoholic beverages'. The home production of food for own consumption and the freely received food have also been valued and recorded separately, in the form of imputed values. In order to circumvent seasonal variation in food expenditure, food consumption by households is monitored over a seven-day period in each of 12 consecutive months. The consumption expenditure in the dataset is the average of this weekly consumption across these 12 months (Department of Census and Statistics, 2011). For the purpose of analysis, the weekly food consumption expenditure data were converted to monthly equivalents.

Non-food expenditure

As given in Table 3.5, the expenditure on non-food items covers 8 sub-groups. In the dataset, the expenditure on non-food items is given in different reference periods, varying from one month to twelve months. The non-food expenditure values are

Table 3.5 Household non-food expenditure sub-categories

	Expenditure Category	Description
1	Housing, household services, and durable and non-durable household goods (Housing)	Refers to the average payment of rent, mortgages, water bills and taxes per month. For owner-occupied houses, an estimated rent has been included. Expenditure on wages to servants, laundry charges and grinding charges, expenditure on washing soap and powder, insecticides for household use, and expenses on household appliances and utensils, furniture, purchase of motor vehicles, bicycles, etc. in the past six months and twelve months
2	Fuel and lighting	Average monthly expenditure towards electricity, gas, other fuel and related items such as bulbs
3	Clothing, textiles and footwear	Expenditure on purchase of clothing, clothing materials, tailoring charges, shoes, bags, umbrellas, etc. and repair charges
4	Health and personal care	Expenses on practitioner's fee, medicines, pharmaceutical products, hospital charges and payments to medical laboratories for test analysis and expenses on toiletries, cosmetics and beauty services in the past month
5	Transport and communication	Expenditure on public and private transport and maintenance of private vehicles including annual revenue license and insurance, and expenses towards telephone, internet and postal charges
6	Education	Education fees for schooling and higher education, school textbooks and stationery
7	Recreation and entertainment	Expenditure towards cinema, books, newspapers, gambling, pilgrimages, sports, pets, toys and maintenance of televisions and other audio-visual equipment
8	Other expenses	Expenditure on provident funds, trade union contributions, insurance, payment of debts, income tax, other savings, weddings and funerals, donations, and other social activities

Source Author's compilation based on Department of Census and Statistics (1993, 2003, 2011, 2014, 2017)

Table 3.6 Household expenditure on non-food items, by survey year (as a percentage of total non-food expenditure)

Expenditure category	1990/1991	2002	2009/2010	2012/2013	2016
(1)	(2)	(3)	(4)	(5)	(6)
Housing	29.6	29.1	24.9	23.3	26.5
Fuel and lighting	11.8	7.6	7.1	6.8	4.9
Clothing, textiles and footwear	10.4	5.3	5.0	4.6	4.4
Health and personal care	9.7	8.0	7.9	8.5	7.0
Transport and communication	12.6	12.7	17	16.7	15.5
Education	4.3	4.3	5.6	5.6	5.8
Recreation and entertainment	2.2	2.0	2.2	2.0	2.5
Alcohol and tobacco	9.4	4.2	3.7	2.7	2.9
Other expenses	10.0	26.8	26.6	29.8	30.5
Total non-food expenditure	100	100	100	100	100

Source Author's compilation based on Department of Census and Statistics (1993, 2003, 2011, 2014, 2017)

converted to monthly data to match with expenditure on food for the empirical analysis of this study.

Table 3.6 summarises the household expenditure patterns on non-food items in the five survey years. While up to 2012/2013, the highest budget share of non-food items were spent on housing, in 2016, the other expenses dominated the household non-food budget, followed by expenditure on housing. Overall, Sri Lankan households allocate the lowest non-food budget share on recreation and entertainment. While household budget shares on housing, and health and personal care have seen a minor reduction during 1990/1991 and 2016, the budget share allocation on fuel and lighting, clothing, and alcohol and tobacco has seen a notable drop over the years. On the other hand, household non-food budget share on transport and communication, education and recreation and entertainment has seen a minor growth during the period while the other expenses category experienced a substantial growth.

Limitations in expenditure data: measurement errors

As the empirical results of this book are predominantly based on household expenditure data, it is worthwhile considering the accuracy of the expenditure data collected via the HIES surveys. Several measurement problems related to household expenditure data have been identified (Chesher & Schluter, 2002; Deaton & Grosh, 2000). For example, research suggests that there may be recall errors in the household survey data in which households are asked to recall their past spending habits. Diary data, used in this study, has been consistently identified as superior to recall data, particularly in relation to expenditure on food (Battistin, 2003; Gibson & Kim, 2007). Ahmed et al. (2006) showed that, although superior to recall data, diary data also appear to have measurement errors, that are mostly associated with "diary fatigue" or "diary exhaustion". The result is a decrease in expenditure records being kept between the first and second week of the diary. In addition to the between week variations, the responses appear to be considerably larger for the earlier days of the week. This can result in an under-reporting of consumption data (Ahmed et al., 2006; Deaton & Grosh, 2000). Although these problems associated with diary data are not entirely resolved, the best practice that has been adopted to minimise these errors when collecting household survey data is to have a relatively short diary period of a week or two (Deaton & Grosh, 2000). In the case of the Sri Lankan HIES, the reference period of the diary used for food items is one week (Department of Census and Statistics, 2011); hence the potential errors associated with use of the diary method can be expected to be at a marginal level.

3.3.3 Poverty Status

As discussed in Sect. 2.4.2, one of the most commonly used poverty indicators for the purpose of policy formulation is the poverty headcount index (HI). It is the percentage of the population whose per capita monthly household expenditure is less than the OPL. The foundation value of the OPL in Sri Lanka is SLR 1,423 and was determined by the DCS according to the 'cost of basic need method' on food and non-food consumption, based on the 2001/2002 HIES series. The OPL is an absolute poverty line, as it is based on 2030 kilocalories, the recommended daily per capita food energy requirement for Sri Lanka. The OPL for the survey years before/after 2002 is obtained by deflating/inflating the base year OPL, using the Colombo Consumer Price Index (CCPI). As such, the poverty line, in real terms, has remained unchanged over time, and hence can be used to compare the changes in poverty status in the country, over time. The OPL used in the 1990/1991 HIES is SLR. 475; in the 2009/2010 HIES this rose to SLR. 3,028 and by 2016 it further rose to SLR. 4,166 (Department of Census and Statistics, 2004, 2017).

Figure 3.3 illustrates the trends in the poverty HI during 1990/1991 and 2016 by sector. As can be seen, the overall head-count index in the country has seen a notable reduction during the period from 26.1% in 1990/1991 to 4.1% in 2016. This reduction in poverty levels in the country is attributed to various poverty reduction strategies and policies undertaken by the respective governments during this period.

When considering the sectoral poverty rates, the urban sector has the lowest head-count index during the period between 1990/1991 and 2016. While in the early 1990s the rural sector had a much higher poverty head-count index than that of the estate sector, since the early 2000s the poverty head-count index in the estate sector surpassed that of the rural sector. Currently, the estate sector records the highest poverty rates of the country, almost twice that of the rural sector and over four times that of the urban sector. Nevertheless, given that the majority of the population in the country (in 2016, about 78%) still lives in the rural sector, the highest number of poor people is seen in the rural sector. This is why it is often said that the poverty in Sri Lanka is predominantly a rural phenomenon (Gunetilleke & Senanayake, 2004).

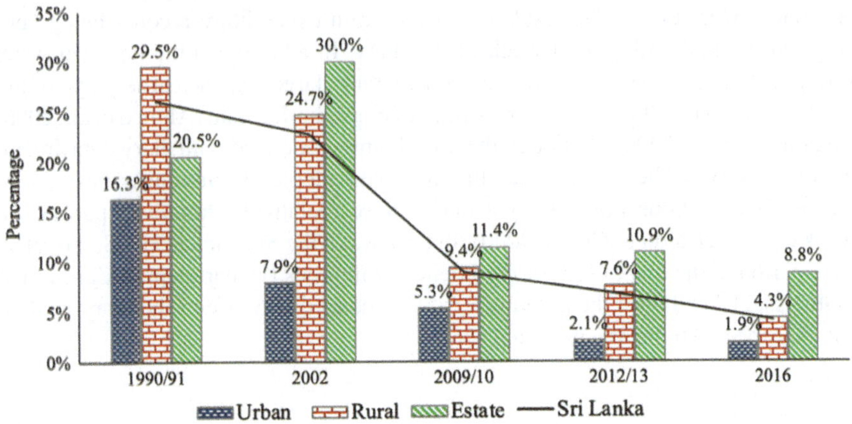

Fig. 3.3 Trends in poverty head-count index, by sector and survey year. *Source* Author's compilation based on Department of Census and Statistics (1993, 2003, 2011, 2014, 2017)

3.4 Concluding Remarks

The purpose of this chapter was to present a comprehensive analysis of the five HIES series, namely 1990/1991, 2001/2002, 2009/2010, 2012/2013, and 2016. The chapter presented an contextual overview of Sri Lanka followed by a comprehensive analysis of the HIES datasets used in this book. It included a detailed analysis of the demographic characteristics of the households and household expenditure, in terms of food and non-food items, and the poverty status of the country, measured using the poverty head-count index. The analysis highlights a number of key features about the socio-economic and demographic characteristics of households in Sri Lanka.

- Household size has declined over the years.
- Household expenditure patterns, in terms of food and non-food expenditure, have seen a considerable change during the period between 1990/1991 and 2016.
- The analysis revealed that the poverty levels have declined dramatically over the years, although it still remains as one of the major policy concerns in the development agenda of the country. Despite the reduction in overall poverty rates at the national level, there exist substantial disparities in sectoral level poverty statuses.

The next chapter, using the 2016 HIES dataset, examines the economies of scale in consumption of home-grown food and their implications on food security in Sri Lanka.

References

Ahmed, N., Brzozowski, M., & Crossley, T. F. (2006). *Measurement errors in recall food consumption data*. IFS Working Papers. Institute for Fiscal Studies (IFS).

Battistin, E. (2003). *Errors in survey reports of consumption expenditures*. IFS Working Papers. The Institute for Fiscal Studies.

Central Bank of Sri Lanka. (2020). *Central Bank annual report*. Central Bank of Sri Lanka.

Chesher, A., & Schluter, C. (2002). Welfare measurement and measurement error. *The Review of Economic Studies, 69*(2), 357–378. https://doi.org/10.1111/1467-937X.00209

Deaton, A., & Grosh, M. E. (2000). Consumption. In M. E. Grosh & P. Glewwe (Eds.), *Designing household survey questionnaires for developing countries: Lessons from 15 years of the living standards measurement study* (pp. 91–133). World Bank.

Deaton, A., & Muellbauer, J. (1980). *Economics and consumer behavior*. Cambridge University Press.

De Silva, W. I. (2005). Family transition in South Asia: Provision of social services and social protection. *Asia-Pacific Population Journal, 20*(2), 13–46.

Department of Census and Statistics. (1993). *Household income and expenditure survey—1990/91*. Department of Census and Statistics.

Department of Census and Statistics. (2003). *Household income and expenditure survey—2002*. Department of Census and Statistics.

Department of Census and Statistics. (2004). *Official poverty line bulletin*. Department of Census and Statistics.

Department of Census and Statistics. (2011). *Household income and expenditure survey—2009/10*. Department of Census and Statistics.

Department of Census and Statistics. (2012). *Census of population and housing—2012*. Department of Census and Statistics.

Department of Census and Statistics. (2014). *Household income and expenditure survey—2012/13*. Department of Census and Statistics.

Department of Census and Statistics. (2017). *Household income and expenditure survey—2016*. Department of Census and Statistics.

Gibson, J., & Kim, B. (2007). Measurement error in recall surveys and the relationship between household size and food demand. *American Journal of Agricultural Economics, 89*(2), 473–489.

Gunetilleke, N., & Senanayake, D. (2004). *Monetary poverty estimates in Sri Lanka: Selected issues*. http://www.adb.org/publications/monetary-poverty-estimates-sri-lanka-selected-issues

Jayasinghe, M., & Smith, C. (2021). Poverty implications of household headship and food consumption economies of scale: A case study from Sri Lanka. *Social Indicators Research, 155*(1), 157–185.

Ruwanpura, K. N., & Humphries, J. (2004). Mundane heroines: Conflict, Ethnicity, gender, and female headship in Eastern Sri Lanka. *Feminist Economics, 10*(2), 173–205 + 267. https://doi.org/10.1080/1354570042000217766.

United Nations Development Program (UNDP). (2020). *Human development report*. United Nations Development Program.

Chapter 4
Food Security, Home-Grown Food Consumption and Economies of Scale

4.1 Introduction

Small-scale vegetable farming is a common feature of rural life in developing countries through which families can supplement their monetary income by growing their own food. The World Bank estimates that four out of 5 people below the international poverty line live in rural areas (World Bank, 2021). According to the Food and Agricultural Organization (FAO) (2021), about 90% of the world's 570 million farms are small and most are found in the rural areas of the developing world. Many rural households in developing countries whose primary source of living is agriculture possess the knowledge required for food cultivation (Coale & Hoover, 2015; Hassan & Babu, 1991). In addition to farming being the main source of livelihood, these rural households fulfill most of their food and nutritional needs through their own agricultural production. There is a lack of literature on the economies of scale in home-grown food consumption. If the consumption of home-grown food contributes to economies of scale in food consumption, then both poverty estimates and the distribution of poverty across rural and urban areas could be severely biased (Jayasinghe & Smith, 2021; Jayasinghe et al., 2016, 2018; Sabates et al., 2001).

This chapter uses unique diary data on the household consumption of home-grown food in Sri Lanka to empirically verify the extent to which the consumption of home-grown food enables households to reap economies of scale in food consumption. As discussed in Chap. 2, there are three sources of economies of scale in consumption; (1) the sharing of household goods, (2) the bulk purchases, and (3) the increasing returns to scale in home production. Because food production is time-intensive, it seems intuitive that the capacity of large households to grow their own food enables them to realise consumption economies of scale by spending relatively lower per capita costs on maintaining a given material standard of living (Nelson, 1988). In other words, there is a possibility of having economies of scale in consumption of home produced food through increasing returns to scale in home production. In this chapter, this proposition is explored using a theoretical model along with empirical evidence that validates the theoretical model.

M. Jayasinghe, *Poverty, Food Consumption, and Economic Development*,
https://doi.org/10.1007/978-981-16-8743-3_4

Another important question in studying economies of scale in food consumption is the extent to which observed reductions in per capita costs really do reflect lower living costs for larger households. Such reductions in PCE could also reflect a reduction in the overall quantity of food consumed. This is because a decline in per capita food consumption may imply that members in larger households may be left consuming less food than their counterparts in small households. Thus, it is also important to show that *ceteris paribus* conditions hold in the sense that reductions in food spending take place while the overall level of consumption does not decline. This is done by estimating the calories consumed, based on the food quantity consumed, for each household and exploring how the household's decision to consume home-grown food impacts on per capita calories consumed. The findings in this chapter have significant implications for addressing food security and malnutrition amongst the poor (Gray et al., 2014; Kortright & Wakefield, 2011; Taylor & Lovell, 2014).

The chapter is structured as follows. Section 4.2 provides a brief literature review. Section 4.3 presents a utility maximizing model of household behaviour that highlights how the presence of economies of scale in consumption of home-grown and market-purchased food alters food spending patterns as the size of the household changes. Section 4.4 discusses the data and Sect. 4.5 discusses the empirical methodology. Section 4.6 presents the empirical results and discussion. Section 4.7 discusses the implications of home-grown food consumption on food security and Sect. 4.8 provides concluding remarks and policy implications.

4.2 Review of Literature

While there is only limited research that examines the direct relationship between home-grown food consumption and economies of scale in consumption (Jayasinghe et al., 2017), a considerable number of studies in the literature highlight that home-grown food plays an integral part in the food supply of low-income and, particularly, rural-based households (Calvet-Mir et al., 2012; Morton, 2007; Trinh et al., 2003) and thereby contributes to global food security (see, for example, Galhena et al., 2012, 2013; Lal, 2020; Marsh, 1998; Thaman, 1995).

In general, rural-based households are more inclined to grow their own food than their urban counterparts. Morton et al. (2008), for example, find that rural areas have approximately three times more household gardens than urban areas. There may be several reasons why rural households have a greater reliance on home-grown food. Firstly, food production is mainly a rural activity and, therefore, the agriculture-based occupation system in rural areas may provide a greater reservoir of human and cultural capital supporting home gardening (Morton et al., 2008). The formal work arrangements that are common in the urban areas, in contrast, could pose time constraints for home gardening activities. Additionally, the urbanisation process may have also encouraged urban households to have a greater reliance on commercial markets to meet food needs. Secondly, compared to urban households, rural households may often have access to resources such as land, soil and water necessary to support home gardening. For example, Schupp and Sharp (2012) observe a strong positive relationship between home gardening and availability of space, where households located in the countryside, on a farm, and/or in a free-standing home appear more often to grow their own food. Thirdly, in rural areas there may be higher levels of social networks to exchange extra harvest. For example, Morton et al. (2008) reveal that the norm of giving food to and receiving food from kin, friends and neighbours is stronger in the rural households than in urban households; this reciprocity plays a significant role in food provisioning in rural areas. Urban households, on the other hand, may face more restrictive public regulations relating to the sale and/or exchange of excess produce compared with those in rural areas.

In the relationship between household income and the tendency to grow their own food, a significant number of studies in the literature suggest that home gardening is primarily motivated by economic reasons and that it may provide important income in-kind for low-income households (Campbell et al., 1993; Mingione, 1991; Schupp & Sharp, 2012). Traditionally, food grown at home for owner consumption has been generally seen as a means of augmenting household income and/or reducing spending on market inputs of food items. This is because households, particularly poor households in rural areas, appear to grow food at home to reduce the need for income (to purchase food from the market) or to generate extra income by selling the additional harvest (Brown & Kulcsar, 2001; Torquebiau, 1992). The income earned by selling home-garden products, and the savings generated from consuming home-grown food instead of market-purchased food, may increase the amount of disposable income that can be used to purchase other domestic products. For example, Iannotti et al. (2009) and Vasey (1985), using data from Nepal, Cambodia and Papua New Guinea, showed that the income generated from selling home-grown products enables these households to purchase additional food items as well as to spend on other activities such as education and other household services and even enables households to accumulate savings funds. Growing food at home has been regarded as less-cost intensive and requiring fewer resources and investments than purchasing food, particularly for resource-poor households.

In poor countries, home gardens have been used as a strategy to increase food security and to overcome the hunger and malnutrition of poor and marginalised populations (Buchmann, 2009; Niñez, 1985). This has been a popular and common source of food supply in many countries around the world. Growing food at home for own consumption has been a long-standing practice among households in Sri Lanka for centuries (Jacow & Alles, 1987). Although more recent urbanisation and its related phenomena have had an adverse impact on the tradition of home gardening, there is a strong government focus, through several national policies, on promoting home gardening in the country, over recent years (Central Bank of Sri Lanka, 2013).

Lal (2020) suggests that home gardening can play an important role in advancing food and nutritional security during and after crisis situations, such as the COVID-19 pandemic. For example, the COVID-19 pandemic has aggravated food insecurity in urban centers because of the disruption in the food supply chain, aggravation of the physical and economic barriers that restrict access to food, and the catastrophic increase in food waste because of labour shortages. Logistically, home-grown food provides easy day-to-day access to fresh vegetables and fruits, leading to enriched and balanced diets by supplementing proteins, vitamins, and minerals (Galhena et al., 2013). Marsh (1998) found that households who have access to home gardens obtain more than 50% of the vegetables, fruits, tubers, and yams from their garden. Thaman (1995) argues that the formal promotion, expansion and improvement of small-scale urban food gardening are direct and economically, socially, technologically and nutritionally appropriate means of bringing about sustainable national development and promoting food security.

The next section discusses the theoretical model used in this chapter to analyse the household's decision to consume home-grown food or food purchased from the market. This model is adapted from Jayasinghe et al. (2017) and is a modified version of the household production model presented in Vernon (2005).

4.3 A Theoretical Model of Economies of Scale

The theoretical model relies on the assumption that home-grown food is a labour-intensive substitute for market-purchased food. For example, households can choose to devote labour to grow their vegetables or buy them from the market. Larger households endowed with more labour will therefore choose to grow more food at home, thereby reducing expenditure on market-purchased food. Consider a household comprised of n identical individuals, who derive utility from three goods, namely market-purchased food (the quantity of x_1), home-grown food (the quantity of x_2) and non-food items (the quantity of x_3). Therefore, the per capita consumption of each good is (x_1/n), (x_2/n) and (x_3/n), respectively. The presence of economies of scale is represented by the parameter ϕ_i (i = 1, 2, 3) which is a function of the household size n, which we denote as $\phi_i(n)$. Then the household utility function U is:

$$U = n \cdot u\left(\frac{x_1}{\phi_1(n)}, \frac{x_2}{\phi_2(n)}, \frac{x_3}{\phi_3(n)}\right) \tag{4.1}$$

Economies of scale may be present in each of these three goods. Economies of scale in market-purchased food, $\phi_1(n)$, may arise if larger households spend less per capita expenditure on purchased food ingredients to consume the same amount of food as smaller households. This may occur as a result of the larger households being able to save more from bulk purchases, quantity discounts, and substitution of home-produced meals for expensive food-away-from-home (Nelson, 1988). Economies of scale in home-grown food, $\phi_2(n)$, may occur as a result of increasing returns to scale in agricultural production (Paul & Nehring, 2005). Economies of scale arising from the consumption of non-food items, $\phi_3(n)$, may occur as a result of the joint consumption of household public goods (Nelson, 1988). The scale of these consumption economies is given by:

$$\phi_i(n) = n^{(1-\sigma_i)} \quad i = 1, 2, 3 \tag{4.2}$$

where $0 \leq \sigma_i \leq 1$ is the scale elasticity of the ith commodity within the household. If $\sigma_i = 0$ then $\phi_i(n) = n$. This implies that the good is a private good that cannot be shared and must be replaced if all members in the household are to enjoy the good to the same degree as a single person household.[1] If $\sigma_i = 1$ then $\phi_i(n) = 1$, implying that the good is a pure public good that can be enjoyed by any number of household members, without diminishing the enjoyment by others in the household. The scale elasticity σ_i is derived by taking the log of both sides of Eq. (4.2).[2]

$$\ln\phi_i(n) = 1 - \sigma_i \ln(n) \tag{4.3}$$

$$\sigma_i = 1 - \frac{\partial \ln\phi_i(n)}{\partial \ln n} \tag{4.4}$$

The budget constraint relating to consumption and prices of two food items, non-food, and household income (I), is given by

$$\sum_{i=1}^{3} P_i x_i = I \tag{4.5}$$

The household problem is to maximise the utility function (4.1) subject to the budget constraint (4.5). After taking the first order conditions with respect to x_1 and x_2, and using them to eliminate the Lagrangian multiplier the following condition is provided for the relationship between prices of purchased food (x_1) and home-grown food (x_2).

$$\frac{P_1}{P_2} = \frac{\partial x_2}{\partial x_1} \cdot \frac{\partial(x_1/\phi_1(n))}{\partial(x_2/\phi_2(n))} \tag{4.6}$$

The Marginal Rate of Substitution (MRS) between market-purchased food (x_1) and home-grown food (x_2) is given by[3]

$$MRS_{2,1} = \frac{\partial x_2}{\partial x_1} = \frac{P_1 \partial(x_2/\phi_2(n))}{P_2 \partial(x_1/\phi_2(n))} \tag{4.7}$$

The MRS between the two commodities depends on the prices and magnitude of the economies of scale for each commodity.

[1] In other words, this is the same as (x_i/n).

[2] A simple analytical solution to isolate σ_i produces undefined results, thus requiring differentiation.

[3] This type of MRS can also be calculated to measure substitutability between (i) purchased food and non-food items and (ii) home-grown food and non-food items, but we are not interested in these alternative measures as it is not the focus of this study.

There will be three different possibilities for the MRS, which we discuss in the following Lemma.

Lemma

1. *If $MRS_{2,1} = 1$, then there is a one-to one substitution between x_1 and x_2. This means that the household prefers home-grown food and market-purchased food equally, even though the consumption scale economies of these goods are different.*
2. *If $MRS_{2,1} > 1$, then a household would be willing to forego more than a unit of market-purchased food to substitute a unit of home-grown food. This is possible if*

either

(i) $P_1 > P_2$, *ceteris paribus*

or

(ii) *economies of scale is higher in purchased food i.e. $\phi_1(n) > \phi_2(n)$, so that the household needs less of purchased food to compensate for the same quantity of forgone home-grown food $[\partial(x_1/\phi_1(n)) < \partial(x_2/\phi_2(n))]$.*

3. *If $MRS_{2,1} < 1$ then a household would be willing to forego less than a unit of market-purchased food to substitute a unit* of home-grown food. *This is possible if*

either

(i) $P_2 > P_1$, *ceteris paribus*

or

(ii) *economies of scale is higher in home-grown food i.e. $\phi_2(n) > \phi_1(n)$, so that the household needs less of home-grown food to compensate for the same quantity of forgone purchased food $[\partial(x_2/\phi_2(n)) < \partial(x_1/\phi_1(n))]$.*

Our empirical strategy will focus on verifying condition (*ii*) of the 3rd option where $\phi_2(n) > \phi_1(n)$. This implies that large households need less home-grown food to compensate for the same quantity of forgone market-purchased food.

To see how the presence of economies of scale influences the relationship between household size and food spending, we derive the per capita demand function. Households maximise the utility function (4.1) subject to the budget constraint (4.5) in terms of effective prices and quantities, by including economies of scale in consumption. The effective price $\left(P_i^*\right)$ and quantity $\left(x_i^*\right)$ of a particular commodity refer to the price and quantity adjusted for economies of scale of consumption of that commodity. The effective quantities (x_i^*) and the effective prices $\left(P_i^*\right)$ are given by,

$$x_i^* = \frac{x_i}{[\phi_1(n)\phi_2(n)\phi_3(n)]} \quad i = 1, 2, 3 \tag{4.8}$$

$$p_i^* = \frac{P_i\,[\phi_1(n)\phi_2(n)\phi_3(n)]}{n} \quad i = 1, 2, 3 \tag{4.9}$$

The effective budget constraint (I^*), adjusted for economies of scale from the consumption of all three commodities is given by

$$I^* = \frac{[\phi_1(n)\phi_2(n)\phi_3(n)]}{n} \tag{4.10}$$

The household consumption problem in terms of effective prices and quantities can be presented in the form of a demand function as follows.

$$x_i^* = g\left[P_1^*, P_2^*, P_3^*, I^*\right] \tag{4.11}$$

To show how the demand for x_1 and x_2 is affected by n, we derive a per capita demand function, which features the elasticity of demand with respect to household size. We multiply both sides of (4.11) by $\frac{[\phi_1(n)\phi_2(n)\phi_3(n)]}{n}$ and use (4.8) to give,

$$\begin{aligned} &\frac{x_i}{[\phi_1(n)\phi_2(n)\phi_3(n)\cdot n]}\cdot\frac{[\phi_1(n)\phi_2(n)\phi_3(n)]}{n} = \frac{[\phi_1(n)\phi_2(n)\phi_3(n)]}{n}\cdot g[P_1^*, P_2^*, P_3^* I^*] = \\ &\frac{[\phi_1(n)\phi_2(n)\phi_3(n)]}{n}\cdot g\left\{\frac{[\phi_1(n)\phi_2(n)\phi_3(n)]}{n}[P_1, P_2, P_3, I]\right\} \end{aligned} \tag{4.12}$$

where we have also used (4.9) and (4.10). Since the demand functions are homogenous of degree zero, we have

$$g\{\frac{[\phi_1(n)\phi_2(n)\phi_3(n)]}{n}\cdot[P_1, P_2, P_3, I]\} = g[P_1, P_2, P_3, I] \tag{4.13}$$

That is

$$\frac{x_i}{n} = \frac{[\phi_1(n)\phi_2(n)\phi_3(n)]}{n}\cdot g[P_1, P_2, P_3, I] \tag{4.14}$$

Taking the natural logarithm of Eq. (4.14), the per capita demand function takes the form,

$$\ln\frac{x_i}{n} = \ln\phi_1(n) + \ln\phi_2(n) + \ln\phi_3(n) - \ln n + \ln\{g[P_1, P_2, P_3, I]\} \tag{4.15}$$

The negative term, ($-\ln n$), in Eq. (4.15) suggests that the per capita demand for good i declines due to economies of scale as household size grows.

To verify this, a reduced form of Eq. (4.15) will be estimated using the data from the HIES 2016, discussed in detail in Chap. 3. The (log of) per capita expenditure on purchased food and home-grown food will be used as the dependent variables, while the (log of) household size will be used as one of the independent variables.

4.4 Data: Household Income and Expenditure Survey (HIES), 2016

As discussed in Chap. 3, the 2016 HIES full sample comprises 21,756 households. The removal of the outliers, including the households with more than 10 members resulted in a sample of 17,000, which is used for the empirical analyses of this chapter. Table 4.1 presents summary statistics of the variables used in this chapter. As can be seen, 3/4th of households' members in the sample are adults across all sectors. The estate sector households consume the highest per capita calories per month, followed by the rural and urban sector residents. The higher per capita calorie consumption in the rural and the estate sectors reflects the need for higher energy intake due to the labour-intensive nature of their work (Mahadevan & Hoang, 2015). These results are consistent with other studies of calorie consumption in Sri Lanka (Ministry of Finance & Planning of Sri Lanka, 2012).

The data also reveals that the proportion of households with employed heads is significantly lower in the estate sector (0.38) compared to households in the urban (0.54) and rural sectors (0.51). A higher proportion of rural residents own agricultural lands (0.89) followed by the urban households (0.79). The estate sector has the lowest proportion of households who own agricultural lands. When considering the monthly household expenditure on home-grown food, as can be seen in Table 4.1, rural households spend the most followed by the urban and estate households. On average, urban households spend more on purchased food than households in the rural and estate sectors.

Since the empirical analysis of this chapter is based on the data on home-grown food consumption, it is worthwhile discussing the nature and accuracy of this data in detail. The HIES record data on a total of 249 food and beverages items that

Table 4.1 Summary statistics

	National	Urban	Rural	Estate
Proportion of adults	0.73 (0.21)	0.74 (0.21)	0.75 (0.21)	0.73 (0.21)
Per capita calorie consumption (per month)	55,842 (20,543)	53,367 (19,587)	56,632 (21,261)	59,837 (20,876)
Proportion of employed	0.52 (0.53)	0.54 (0.49)	0.51 (0.48)	0.38 (0.42)
Proportion of agricultural landowners	0.82 (0.41)	0.79 (0.38)	0.89 (0.27)	0.34 (0.48)
Expenditure on home-grown food (in Sri Lankan rupees)	828 (1,341)	321 (452)	1,094 (1,321)	311 (342)
Expenditure on purchased food (in Sri Lankan rupees)	18,286 (8,926)	23,787 (9,873)	17,089 (7,452)	16,579 (6,546)

Notes Standard deviations are given in parenthesis

belong to 18 sub-categories. For each of these food items, an imputed value of the home-grown food and the value of the food purchased from the market are recorded separately. Mostly, food categories, such as cereals, pulses, vegetables, yams and other, coconuts, condiments, fruits, eggs and milk recorded values for the home-grown component. As these items are not purchased, the values of these items are estimated by the household.

The data collection process includes measures to double-check the accuracy of these imputed values (Department of Census & Statistics, 2009). At the beginning of the reference week, the department representative distributes the diary sheet to the household for the household to record the expenditure data related to their food consumption on the previous day. As such, the expected recall period of the food data is limited to one day. On day three, the representative makes another visit to ensure the household is completing the diary sheet properly. Finally, at the end of 7 days, the representative makes a final visit to the household to collect the diary sheet and conduct a short interview in order to ensure the data entered is accurate. During the interview, particular attention is paid to the accuracy of quantities of home-grown food and their values, as those are not obtained through market purchases and hence may pose serious issues on measurement errors in food expenditure data if not reported accurately. The representative is instructed to go through each and every food expenditure sub-category and check for any discrepancies in the value of home-grown/freely received food based on the quantities households have entered. If there is any discrepancy, the representative will correct the values based on the prices of those foods in the respective area where the household lives. A one-week diary period is considered relatively short and hence minimizes the possibility of diary fatigue and underreporting (Ahmed et al., 2006; Deaton & Grosh, 2000). In general, diary data are also considered more accurate than consumption data based on recall surveys (Battistin, 2003; Gibson & Kim, 2007). Jayasinghe et al. (2017) found that the distribution of imputed values of home-grown food overlaps with the distribution of market prices using 2009/2010 HIES data for Sri Lanka. This observation provides reasonable assurance about the accuracy of the imputed values.

The estimation results of this chapter have been adjusted for inter-district or regional differences in food prices, a common issue encountered in nation-wide cross-sectional data (see Fan & Wei, 2006; Banerjee & Duflo, 2007), using the Spatial Price Index (SPI) for 2016 calculated by the DCS (Department of Census and Statistics, 2017). This ensures that results are not biased by regional price differences in food items.

4.5 Empirical Methodology

This section presents the methodology used for the empirical analysis of this chapter.

4.5.1 Economies of Scale in Home-Grown and Market-Purchased Food Consumption: Model Specification

For the empirical analysis, two separate multiple regression models, log of per capita expenditure on home-grown $\ln(hg)$ and market-purchased food $\ln(mp)$ will be estimated using the Seemingly Unrelated Regression (SUR) approach. This is because, the SUR approach allows the comparison of coefficients of two regression models with different dependent variables (Srivastava & Giles, 1987). This enables us to compare the relationship between household size and per capita expenditure on home-grown and purchased food, in terms of the magnitude as well as the direction. The regression specification of the log of per capita household spending on home-grown food $\ln(hg)$, in the form of a reduced form of Eq. 4.15 is given by

$$\ln(hg) = \beta_0 + \beta_1 \ln(n) + \beta_2 \ln() + \beta_3 (\ln(I))^2 + \beta_4 LD + \beta_5 ED + \boldsymbol{\delta}' \boldsymbol{K} + \varepsilon \tag{4.16}$$

where $\ln(n)$ is the log of household size, $\ln(I)$ is log of household income, $(\ln(I))^2$ is the square of log of household income, LD is a dummy variable for ownership of agricultural land, ED is a dummy variable for households with an employed head, and $\boldsymbol{K}$ is a vector of demographic indicators i.e. proportion of adults, gender of the headship (female = 1 and male = 0) and two location dummy variables for rural and estate sector taking the urban sector as the base, and the age and education level of the head of the household with $\boldsymbol{\delta}'$ being the corresponding coefficient vector,[4] and ε is a random error term. The log of per capita demand for purchased food $\ln(mp)$ is also estimated from a specification similar to Eq. (4.16), taking the $\ln(mp)$ as the dependent variable.[5]

If economies of scale exist, the β_1 coefficients are expected to have a negative effect on per capita demand for home-grown and purchased food. In terms of the magnitude, if the negative relationship is found to be larger in per capita demand for home-grown food, it confirms the presence of greater economies of scale in home-grown food to that of purchased food. Additionally, β_2 and β_5 coefficients are expected to have a negative relationship with $\ln(hg)$ and a positive relationship with $\ln(mp)$. This is because the higher income levels and 'employed' status of the head of the household may encourage households to consume more purchased food and

[4] The household income we use in this study is the actual household income (not household expenditure used as proxy measure of household income). This household income consists of all forms of income i.e. income from salaries and wages, income from agricultural activities, income from non-agricultural activities, and other cash receipts such as pension, disability relief payments, government transfer payments, remittances and dividends.

[5] In terms of multicollinearity, the correlation coefficients between the independent variables were relatively low. The highest (absolute) correlation coefficient of 0.534 was reported between the rural and estate dummy variables.

less home-grown food. β_4 coefficient is expected to have a positive relationship with $\ln(hg)$ and a negative relationship with $\ln(mp)$.

The coefficient of estate dummy variable listed under a vector of demographic indicator $\boldsymbol{K}$ is also expected to have a negative relationship with $\ln(hg)$ and a positive relationship with $\ln(mp)$. This is due to time and space constrains and differences in the nature of work, the estate sector households may be less able to produce home-grown food than their rural sector counterparts. The expected sign of the variable on education level, listed under $\boldsymbol{K}$ could be either positive or negative. For example, when the education level of the head of the household improves, it may provide them with greater opportunities to find employment in the formal sector. Hence the opportunity cost of growing food for the own consumption of the educated households may be higher. This would result in less home-grown food consumption and hence the coefficient is negative. Alternatively, when the education level of the head of the household improves it may increase the awareness of the health benefits of consuming home-grown food, resulting in an increase in home-grown food consumption, hence the coefficient is positive.

4.5.2 Link Between Home-Grown Food Consumption and Per Capita Calorie Intake

Another important aspect to investigate is how a greater reliance on home-grown food may impact the calorie intake of households. An important *ceteris paribus* condition for observing economies of scale is that households consuming more home-grown food are not in fact consuming less food per capita. This is particularly important in light of previous studies, which found that larger households tend to consume low per capita calories (Subramanian & Deaton, 1996). As such, it is important to rule out the possibility that a greater reliance on home-grown food may result in decline in overall household calorie consumption.

For this purpose, the per capita calorie consumption data were calculated using tables of food composition statistics derived from Perera et al. (1979). These tables are still widely used in many recent reports (Jayawardena et al., 2014; Nestel et al., 2003) as they cover a wide variety of unique Sri Lankan food types that are not found in other recent international food composition reports. Per capita calorie consumption was calculated by multiplying food quantities by the relevant calorie value provided by these tables. For example, a household that reported consuming 250 g of prawns receives 222.5 cal (100 g of prawns is equivalent to 89 cal).

To check whether there exists a non-negative relationship between per capita calorie consumption and the household's decision to consume more home-grown food, the log of per capita calorie consumption $\ln(c_{pc})$ is regressed on the set of independent variables as specified in Eq. (4.17).

$$\ln(c_{pc}) = \beta_0 + \beta_1\left(\frac{E_{hg}}{E_f}\right) + \beta_2\ln(n) + \beta_3(\ln(n))^2 + \beta_4\ln(I) + \beta_5 LD + \beta_6 ED + \delta' \boldsymbol{K} + \varepsilon \tag{4.17}$$

where (E_{hg}/E_f) is the ratio of expenditure on home-grown food to total food expenditure, where $\ln(n)$ is the log of household size, $(\ln(n))^2$ is the square of log of household size, $\ln(I)$ is log of household income, LD is a dummy variable for ownership of agricultural land, ED is a dummy variable for households with an employed head, and $\boldsymbol{K}$ is a vector of demographic indicators i.e. proportion of adults, gender of the household head (female = 1 and male = 0) and two location dummy variables for the rural and estate sectors taking the urban sector as the base, and the age and education level of the head of the household with $\boldsymbol{\delta}'$ being the corresponding coefficient vector. ε is a random error term. A high E_{hg}/E_f value indicates that a larger proportion of the household's food is sourced through home-grown food.

The β_1 coefficient is expected to be positive, which indicates that households that rely more on home-grown food tend to consume more per capita calories. Consistent with previous results found in the literature where larger households tend to consume low per capita calories, β_2 coefficient is expected to be negative (Subramanian & Deaton, 1996) and β_3 coefficient is expected to be positive. The rural and estate sector dummy variables, listed in the vector of demographic variables $\boldsymbol{\delta}'$, are expected to be positive as these households require higher calorie intake due to the nature of their work, which involves strenuous physical activities.

4.6 Results and Discussion

This section presents the results and discussion on the two model specifications presented in Sect. 4.5.

4.6.1 Economies of Scale in Home-Grown and Purchased Food Consumption

The regression results for demand of home-grown and purchased food based on Eq. (4.16) are presented in Table 4.2. As can be seen in Columns (2) and (3), the majority of the estimated coefficients are statistically significant at the 5 or 10% level of significance. The variables, proportion of adults in the household, age and education of the household head are not statistically significant in either model, while the estate dummy variable in the purchased food model is not significant. The coefficients of main interest in this Table are the β_1 coefficients for the two models. Note that the log–log formulation of the two regression Equations allows us to interpret the β_1 coefficients in Columns (2) and (3) as household size elasticity of per capita consumption of home-grown and purchased food, respectively. As anticipated in Eq. (4.15) of our theoretical model, both β_1 coefficients are negative, suggesting that per capita demand for both home-grown and purchased food declines with household size due to potential economies of scale. For example, for a 10%

Table 4.2 Results on per capita demand for home-grown food and purchased food

(1)	Equation (4.16) with Log of per capita home-grown food **ln(*hg*)** as dep var (2)		Equation (4.16) with Log of per capita purchase food **ln(m*p*)** as dep var (3)	
	Coefficient	SE	Coefficient	SE
Constant (β_0)	11.432*	2.102	5.864*	0.892
Log of household size(β_1)	−0.921*	0.087	−0.482*	0.051
Log of household income (β_2)	−1.201*	0.210	−0.197*	0.021
Square of log of household income (β_3)	0.088*	0.041	0.024*	0.012
Ownership of agricultural land (β_4)	1.824*	0.564	−0.032*	0.004
Employed dummy (β_5)	−0.080*	0.028	0.021*	0.002
Vector of demographic variables ($\boldsymbol{\delta}_i$)				
Proportion of adults	0.161	0.171	0.051	0.080
Rural dummy	0.087*	0.021	−0.190*	0.082
Estate dummy	−0.928*	0.290	0.061	0.120
Female dummy	−0.170**	0.095	−0.121*	0.038
Age	0.011	0.021	0.001	0.021
Education	0.059	0.091	−0.010	0.056
R squared	0.1721		0.2082	
Number of observations	17,000		17,000	

Notes To enable the compassion of coefficients in Columns (2) and (3), the 2 models have been estimated using the Seemingly Unrelated Regression approach. * $p < 0.05$ and ** $p < 0.10$

increase in household size, we expect about a 9.2 and 4.8% decline in the demand for home-grown and purchased food, respectively. More importantly, a larger reduction in the demand for home-grown food than that of purchased food implies that the potential economies of scale associated with home-grown food are larger than those for purchased food.

The estimation results for other variables are in line with the findings in the literature. Confirming Engel's Law, increase in household income (β_2) is associated with declines in the per capita spending on home-grown food and purchased food. The negative relationship between home-grown food and income is also consistent with the findings of Brown and Kulcsar (2001), Campbell et al. (1993), Mingione (1991) and Schupp and Sharp (2012). They argue that growing food at home may be primarily motivated by economic necessity and thus it is an adaptive strategy, mainly, for lower income households. The ownership of agricultural lands (β_4) is positively associated with the demand for home-grown food while it is negatively associated with the demand for purchased food. This confirms Levitan and Feldman's (1991) argument that households with access to land are more likely to grow food at home. The estimate of (β_5) for home-grown food model is negative and statistically significant, indicating that the 'employed' status of the household head leads to a reduction in the consumption of home-grown food. This may be because the presence of a wage-earning household head reduces the need for reliance on home-grown food. This aligns with some findings in the literature (see, for example, Mahroof & Rafeek, 2004) that suggest the changing patterns of employment opportunities in the formal sectors, such as the garment sector in Sri Lanka reduces the availability of labour supply for rural agriculture.

On the other hand, as seen in a positive and statistically significant (β_5) coefficient in purchased food model, the 'employed' status of the household head results in increased consumption of purchased food.

In the vector of demographic variables $\boldsymbol{\delta}'$, the results indicate that households living in rural areas consume more home-grown food (column 2) and less purchased food (column 3) than their urban counterparts. This observation supports Morton et al. (2008), which shows that growing food at home and exchange of food is a significant source of food provisioning among rural households. The lower access to space and water resources, the higher opportunity cost of labour, engagement in formal occupational base and reliance on the commercial markets to fulfil food requirements in the urban sector may explain the lower tendency for urban households to grow food at home. However, results revealed that estate sector households consume lower home-grown food than urban households. This could reflect limited access to land (as they live in housing schemes provided by the estate owners) and the high opportunity cost of time of home gardening (as the majority of them are estate sector workers) that may generate constraints on the capacity for growing food at home. The occupational base in the estate sector mainly revolves around tea estates and hence, naturally, these households are not trained in more traditional consumption-oriented agricultural activities.

4.6.2 Link Between Home-Grown Food Consumption and Per Capita Calorie Intake

Table 4.3 presents the results on the determinants of household per capita calorie consumption [Eq. (4.17)]. The F statistic with $p < 0.0000$ indicates that the overall utility of the model is very good. Estimated coefficients, except for the household head's education, are statistically significant at the 5% level, and are in line with the finding in the literature with respect to household calorie consumption in developing countries.

When considering the relationship between home-grown food consumption and calories, the results show that those larger households who rely more on home-grown food tend to consume more calories per capita. The positive β_1 coefficient shows that a higher ratio of home-grown to total food expenditure is associated with an 18%

Table 4.3 Results on per capita calorie consumption

Dependent variable	Log of per capita calorie consumption $\ln(c_{pc})$	
	Coefficient	S.E.
Constant (β_0)	9.567*	0.032
Home-grown to total food ratio (β_1)	0.181*	0.034
Log of household size(β_2)	−0.321*	0.043
Square of log of household size(β_3)	−0.042*	0.009
Log of household income (β_4)	0.092*	0.002
Ownership of agricultural land (β_5)	0.041*	0.010
Employed dummy (β_6)	−0.054*	0.007
Vector of demographic variables (δ_i)		
Proportion of adults	−0.054*	0.012
Rural dummy	0.041*	0.002
Estate dummy	0.153*	0.024
Female dummy	−0.062*	0.005
Age	0.010*	0.001
Education	0.003	0.007
R squared	0.1982	
Number of observations	17,000	

Notes * $p < 0.05$ and ** $p < 0.10$

increase in per capita calories consumed by the household.[6] Similar results were found in Jayasinghe et al. (2017). This confirms that the reduction in the per capita consumption of home-grown food that occurs among large households (see β_1 in Table 4.2) is not due to a reduction in calories consumed by the household, but due to the presence of economies of scale in home-grown food consumption.

Similar to the findings of Subramanian and Deaton (1996), the estimation results in Table 4.3 reveal that large households tend to consume fewer per capita calories than smaller households. This is reflected in the negative sign of β_2 coefficient in Table 4.3. In particular, the estimation results suggest that for a 10% increase in household size, the per capita calories consumed fall by 3.2%. Also consistent with Subramanian and Deaton (1996) and Bouis (1994), the results also suggest that the income elasticity of per capita calories consumed is relatively low (see β_4). Households located in rural areas and those that own agricultural land are found to consume more calories per capita than their respective counterparts.

[6] As this variable is in the log linear form, it is interpreted as Exp (0.181) = 1.198.

4.7 Implications of Home-Grown Food Consumption on Food Security

While there are numerous definitions of food security, overall, all these definitions cover three broad dimensions: food availability, accessibility, and adequacy/utilization (FAO, 2008). Food availability signifies the availability of food through domestic production, net imports, food reserves, donations, etc. Accessibility refers to the ability of individuals to obtain food without any physical, social, or economic barriers. Food adequacy/utilisation refers to the utilisation of food through an adequate diet that ensures sufficient energy and nutrient intake. Food sourced from home-gardens plays a critical role in directly contributing to household food security by increasing availability, accessibility, and utilisation of food products. The existing literature suggests that home-grown food is widely used as a remedy to alleviate hunger and malnutrition, particularly in developing countries. Home-grown food also provides a solution to the food and nutrition problem by increasing the availability of affordable food and buffering the household against food shortages. Some studies have found a positive relationship between successful home gardens and households' nutritional status (Ochse, 1937), household food consumption and home food production (Danoesastro, 1980), and household calorie intake and home-grown food consumption (Jayasinghe et al., 2017). For poor and vulnerable households who are unable to afford expensive animal products to fulfil their nutritional needs, the home-grown food fulfils most of their calorie needs in a more affordable manner (Girard et al., 2012; Kortright & Wakefield, 2011; Masset et al., 2012).

Figure 4.1 compares the differences in diets of two-adults and two-children households who rely on home-grown food and purchased food in rural Sri Lanka. Panel A presents the diet of households whose home-grown food budget share as a proportion of total food expenditure is 40% or more, while Panel B presents that of households who spent less than 40% on the same (Panel B). The results also reveal that the former consumes an average of 56,965 cal per capita and the latter consumes 55,010 per capita calories per month. As can be seen in Fig. 4.1, the households who spend more than 40% of their food budget on home-grown food fulfil most of their calorie

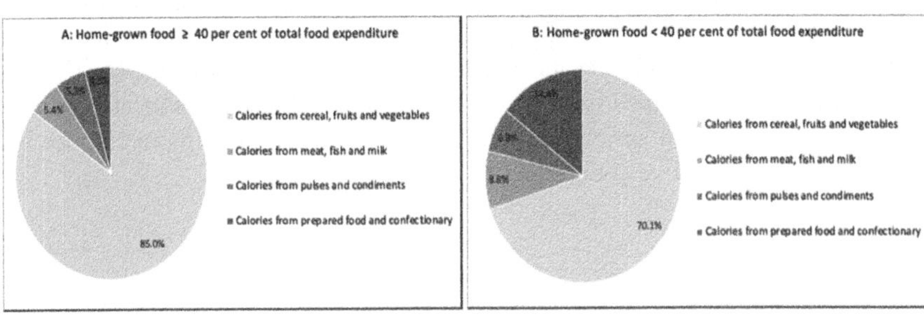

Fig. 4.1 Sources of household calorie intake by major food group

intake through home-grown items, such as cereal, vegetables and fruits. This represents 85% of the total calories for the households in Panel A and is about 15% higher than the calorie intake from cereal, vegetables and fruits in households who spend less than 40% of their food budget on home-grown food (Panel B).

Figure 4.1 also reveals evidence in support of the existence of a substitution effect between calories sourced from cereal, vegetables and fruits and prepared food and confectionary. For example, while households in Panel A consumes only about 4.3% of calorie intake from prepared food and confectionary, households in Panel B source 14.4% of their per capita monthly calorie needs from prepared food and confectionary. However, the results show that calorie intake from protein and dairy sources is about 1.5% lower for households who are relying on home-grown food than those who rely on purchased food. To this end, the analysis in this chapter also supports the argument that household who consume food grown at home achieve better nutritional outcomes, contributing further to household food security.

4.8 Concluding Remarks and Policy Implications

The consumption of home-grown food is a dimension of rural life that has thus far been neglected as a potential source of consumption economies of scale in developing countries. In this study, empirical evidence has been provided to show that scale economies do exist in the consumption of this type of food, which is a time-intensive alternative source of nutrition to market purchased food. The magnitude of these economies of scale in home-grown food is larger than those found in market-purchased food. Moreover, the chapter also provides empirical evidence that consumption of food grown at home is associated with higher per capita calorie consumption. Therefore, those households who are inclined to consume more home-grown food appear to be nutritionally better off compared to those households who tend to consume more purchased food.

These results suggest that one way to enhance food security is to support households, especially those who are vulnerable to poverty and food insecurity, to grow food for their own consumption. Such initiatives may also enhance the nutritional status of poor households. For example, Bangladesh has shown significant success in increasing the availability and consumption of vitamin A-rich food items through national home gardening programs (Talukder et al., 2000). The Sri Lankan government has also identified the importance of supporting the 'backyard' economy among the low-income households to raise their standard of living (Ministry of Finance & Planning of Sri Lanka, 2012).

Home-gardening of food can be encouraged through information campaigns educating households about the socio-economic and health benefits of the consumption of home-grown food and types of food crops that can be cultivated in small plots of land. For those households who do not have access to their own plots, the provision of small alternatives such as containers, trellises and hydroponic gardening equipment could also be considered. The promotion of community seed sharing projects and the distribution of compost bins that provide a cheap and sustainable source of fertilizer are other ways through which policymakers can promote sustainable gardening in developing countries (FAO, 1997). In the wake of increasing rural to urban migration, such initiatives to encourage urban households, with restricted access to cultivatable plots, to grow food for their own consumption is important. Availability of adequate food for household members through home gardens will enhance household resilience and ensure food security in the face of economic crisis.

References

Ahmed, N., Brzozowski, M., & Crossley, T. F. (2006). *Measurement errors in recall food consumption data.*

Banerjee, A. V., & Duflo, E. (2007). The economic lives of the poor. *The Journal of Economic Perspectives, 21*(1), 141–167. https://doi.org/10.1257/089533007780095556

Battistin, E. (2003). *Errors in survey reports of consumption expenditures.*

Bouis, H. E. (1994). The effect of income on demand for food in poor countries: Are our food consumption databases giving us reliable estimates? *Journal of Development Economics, 44*(1), 199–226. https://doi.org/10.1016/0304-3878(94)00012-3

Brown, D. L., & Kulcsar, L. (2001). Household economic behavior in post-socialist rural Hungary. *Rural Sociology, 66*(2), 157–180. https://doi.org/10.1111/j.1549-0831.2001.tb00062.x

Buchmann, C. (2009). Cuban home gardens and their role in social-ecological resilience. *Human Ecology, 37*(6), 705–721. https://doi.org/10.1007/s10745-009-9283-9

Calvet-Mir, L., Gómez-Baggethun, E., & Reyes-García, V. (2012). Beyond food production: Ecosystem services provided by home gardens. A case study in Vall Fosca, Catalan Pyrenees, Northeastern Spain. *Ecological Economics, 74*, 153–160. https://doi.org/10.1016/j.ecolecon.2011.12.011

Campbell, R. R., Spencer, J. C., & Amonker, R. G. (1993). The reported and unreported Missouri Ozarks: Adaptive strategies of the people left behind. In T. A. Lyson & W. W. Falk (Eds.), *In forgotten places: Uneven development in rural America* (pp. 30–52). University Press of Kansas.

Central Bank of Sri Lanka. (2013). *Annual report*. Central Bank of Sri Lanka.

Coale, A. J., & Hoover, E. M. (2015). *Population growth and economic development*. Princeton University Press.

Danoesastro, H. (1980). *The role of home gardens as a source of additional daily income.* Paper presented at the Seminar on the Ecology of Home gardens III.

Deaton, A., & Grosh, M. E. (2000). Consumption. In M. E. Grosh & P. Glewwe (Eds.), *Designing household survey questionnaires for developing countries: Lessons from 15 years of the living standards measurement study* (pp. 91–133). World Bank.

Department of Census and Statistics. (2009). *Enumerator instruction manual-household income and expendirure survey. Sample survey devision.* Department of Census and Statistics.

Department of Census and Statistics. (2017). *Poverty indicators.*

Fan, C. S., & Wei, X. (2006). The law of one price: Evidence from the transitional economy of China. *The Review of Economics and Statistics, 88*(4), 682–697. https://doi.org/10.1162/rest.88.4.682

Food and Agriculture Organization of the United Nations. (1997). *Agriculture food and nutrition for Africa—A resource book for teachers of agriculture* (ISBN 92-5-103820-1). http://www.fao.org/docrep/w0078e/w0078e00.htm

Food and Agriculture Organization of the United Nations. (2021). *Smallholder family farms*. http://www.fao.org/economic/esa/esa-activities/smallholders/en/

Food and Agriculture Organization. (2008). An introduction to the basic concepts of food security. Food security information for action, practical guides. EC—FAO Food Security Programme.

Galhena, D. H., Mikunthan, G., & Maredia, K. M. (2012). Home gardens for enhancing food security in Sri Lanka. *Farming Matters, 28*(2), 12.

Galhena, D. H., Freed, R. & Maredia, K. M. (2013). Home gardens: A promising approach to enhance household food security and wellbeing. *Agriculture & Food Security, 2*(8). https://doi.org/10.1186/2048-7010-2-8

Gibson, J., & Kim, B. (2007). Measurement error in recall surveys and the relationship between household size and food demand. *American Journal of Agricultural Economics, 89*(2), 473–489.

Girard, A. W., Self, J. L., McAuliffe, C., & Olude, O. (2012). The effects of household food production strategies on the health and nutrition outcomes of women and young children: A systematic review. *Paediatric and Perinatal Epidemiology, 26*(s1), 205–222.

Gray, L., Guzman, P., Glowa, K. M., & Drevno, A. G. (2014). Can home gardens scale up into movements for social change? The role of home gardens in providing food security and community change in San Jose, California. *Local Environment, 19*(2), 187–203.

Hassan, R. M., & Babu, S. C. (1991). Measurement and determinants of rural poverty: Household consumption patterns and food poverty in rural Sudan. *Food Policy, 16*(6), 451–460.

Iannotti, L., Cunningham, K., & Ruel, M. (2009). Improving diet quality and micronutrient nutrition: homestead food production in Bangladesh. *IFPRI Discussion Paper 00928*. International Food Policy Research Institute (IFPRI).

Jacow, V. J., & Alles, W. S. (1987). Kandyan gardens of Sri Lanka. *Agroforestry Systems, 5*, 123–137.

Jayasinghe, M., & Smith, C. (2021). Poverty implications of household headship and food consumption economies of scales: A case study from Sri Lanka. *Social Indicators Research, 155*(1), 157–185. https://doi.org/10.1007/s11205-020-02560-z

Jayasinghe, M., Smith, C., Chai, A., & Ratnasiri, S. (2016). The implications of income dependent equivalence scales for measuring poverty in Sri Lanka. *International Journal of Social Economics, 43*(12), 1–17. https://doi.org/10.1108/IJSE-03-2015-0061

Jayasinghe, M., Chai, A., Ratnasiri, S., & Smith, C. (2017). The Power of the vegetable patch: How home-grown food helps large rural households achieve economies of scale and escape poverty. *Food Policy, 73*, 62–74. https://doi.org/10.1016/j.foodpol.2017.09.005

Jayasinghe, M., Ratnasiri, S., Smith, C., & Chai, A. (2018). Domestic technology, consumption economies of scale and poverty: Evidence from Sri Lanka. *Applied Economics, 50*(16), 1777–1789. https://doi.org/10.1080/00036846.2017.1374540

Jayawardena, R., Thennakoon, S., Byrne, N., Soares, M., Katulanda, P., & Hills, A. (2014). Energy and nutrient Intakes among Sri Lankan Adults. *International Archives of Medicine, 7*(1), 34–34. https://doi.org/10.1186/1755-7682-7-34

Kortright, R., & Wakefield, S. (2011). Edible backyards: A qualitative study of household food growing and its contributions to food security. *Agriculture and Human Values, 28*(1), 39–53.

Lal, R. (2020). Home gardening and urban agriculture for advancing food and nutritional security in response to the COVID-19 pandemic. *Food Security, 12*, 871–876. https://doi.org/10.1007/s12571-020-01058-3

Levitan, L., & Feldman, S. (1991). For love or money: Nonmonetary economic arrangements among rural households in Central New York. In D. Clay & H. Schwarzweller (Eds.), *Research in rural sociology and development* (Vol. 5, pp. 149–172). JAI Press.

Mahadevan, R., & Hoang, V. (2015). Is there a link between poverty and food security? *Social Indicators Research, 128*, 179–199. https://doi.org/10.1007/s11205-015-1025-3

Mahroof, A. R. M., & Rafeek, M. I. M. (2004). *Mechanization of paddy harvesting: The economic perspective.*

Marsh, R. (1998). Building on traditional gardening to improve household food security. *Food Nutrition and Agriculture, 22*, 4–14.

Masset, E., Haddad, L., Cornelius, A., & Isaza-Castro, J. (2012). Effectiveness of agricultural interventions that aim to improve nutritional status of children: systematic review. *BMJ, 344*, d8222.

Mingione, E. (1991). *Fragmented societies: A sociology of economic life beyond the market paradigm*. Basil Blackwell Ltd.

Ministry of Finance and Planning of Sri Lanka. (2012). Annual report.

Morton, J. F. (2007). The impact of climate change on smallholder and subsistence agriculture. *Proceedings of the National Academy of Sciences of the United States of America, 104*(50), 19680–19685. https://doi.org/10.1073/pnas.0701855104

Morton, L. W., Bitto, E. A., Oakland, M. J., & Sand, M. (2008). Accessing food resources: Rural and urban patterns of giving and getting food. *Agriculture and Human Values, 25*(1), 107–119.

Nelson, J. A. (1988). Household economies of scale in consumption: Theory and evidence. *Econometrica, 56*(6), 1301–1314.

Nestel, P., Nalubola, R., Sivakaneshan, R., & Wickramasinghe, A. R. (2003). The use of iron-fortified wheat flour to reduce anemia among the estate population in Sri Lanka. *International Journal for Vitamin and Nutrition Research, 74*(1), 35–51. https://doi.org/10.1024/0300-9831.74.1.35

Niñez, V. K. (1985). Working at half-potential: Constructive analysis of home garden programmes in the Lima slums with suggestions for an alternative approach. *Food and Nutrition Bulletin, 7*, 6–13.

Ochse, J. J. (1937). Horticulture and its importance in diet research. *Landbouw, 3*, 202–225.

Paul, C. J. M., & Nehring, R. (2005). Product diversification, production systems, and economic performance in U.S. agricultural production. *Journal of Econometrics, 126*(2), 525–548. https://doi.org/10.1016/j.jeconom.2004.05.012

Perera, W. D. A., Jayasekara, P. M., & Thaha, S. Z. (1979). Tables of food composition for use in Sri Lanka.

Sabates, R., Gould, B. W., & Villarreal, H. J. (2001). Household composition and food expenditures: A cross-country comparison. *Food Policy, 26*(6), 571–586.

Schupp, J. L., & Sharp, J. S. (2012). Exploring the social bases of home gardening. *Agriculture and Human Values, 29*(1), 93–105. https://doi.org/10.1007/s10460-011-9321-2

Srivastava, V. K., & Giles, D. E. A. (1987). *Seemingly unrelated regression equations models*. Marcel Dekker Inc.

Subramanian, S., & Deaton, A. (1996). The demand for food and calories. *Journal of Political Economy, 104*(1), 133–162. https://doi.org/10.1086/262020

Talukder, A., Kiess, L., Huq, N., de Pee, S., Darnton-Hill, I., & Bloem, M. W. (2000). Increasing the production and consumption of vitamin A-rich fruits and vegetables: Lessons learned in taking the Bangladesh homestead gardening programme to a national scale. *Food and Nutrition Bulletin, 21*(2), 165–172.

Taylor, J. R., & Lovell, S. T. (2014). Urban home food gardens in the Global North: Research traditions and future directions. *Agriculture and Human Values, 31*(2), 285–305.

Thaman, R. R. (1995). Urban food gardening in the Pacific Islands: A basis for food security in rapidly urbanising small-island states. *Habitat International, 19*(2), 209–224. https://doi.org/10.1016/0197-3975(94)00067-C

Torquebiau, E. (1992). Are tropical agroforestry home gardens sustainable? *Agriculture, Ecosystems and Environment, 41*(2), 189–207. https://doi.org/10.1016/0167-8809(92)90109-O

Trinh, L. N., Watson, J. W., Hue, N. N., De, N. N., Minh, N. V., Chu, P., & Eyzaguirre, P. B. (2003). Agrobiodiversity conservation and development in Vietnamese home gardens. *Agriculture, Ecosystems and Environment, 97*(1), 317–344. https://doi.org/10.1016/S0167-8809(02)00228-1

Vasey, D. E. (1985). Household gardens and their niche in Port Moresby, Papua New Guinea. *Food and Nutrition Bulletin, 7*(3), 37–43.

Vernon, V. (2005). Food expenditure, food preparation time and household economies of scale. State University of New York.

World Bank. (2021). Poverty overview. https://www.worldbank.org/en/topic/poverty/overview

Chapter 5
Access to Electricity, Technology Adoption and Consumption Economies of Scale

5.1 Introduction

It is widely acknowledged that access to electricity enhances household welfare (Awaworyi Churchill et al., 2020; Fan et al., 2005; Jayasinghe et al., 2021). This is because, as found in a significant body of literature, access to electricity increases household income, considering both farm income (Barnes et al., 2009) and nonfarm income (Cook et al., 2005). However, the literature on the effects of electrification on poverty reduction is still scarce, in particular the specific channels through which poverty reduction may take place has not been adequately examined (Kooijman-van Dijk, 2012). Annemarije (2008) shows that one of the reasons why there is little understanding of the link between electrification and poverty and inequality is because the relationship consists of several steps and many factors influence each of these steps, making the analysis somewhat complicated. One way to explain the link between electricity access and poverty is to explore how domestic technology adoption, facilitated by household electricity access, affects consumption economies of scale and thereby the measurement of poverty.

It is widely agreed that domestic technology adoption, assisted by electricity access has lowered the cost of everyday household tasks, particularly by reducing the amount of labour required in the accomplishment of these tasks (De Vries, 2008; Hall & Khan, 2003; Mokyr, 2000; Vanek, 1978). For example, in a day, cooking food and related tasks, such as gathering firewood, consumes a significant amount of time for females, particularly in the rural areas of developing countries. Use of domestic technologies such as microwave ovens and refrigerators in food preparation, processing, and preservation, and washing machines and clothes dryers in the laundry saves time that can be spent on other tasks such as an income-earning activity or even on leisure (Hossain, 2010; Wickramasinghe, 2011). Over time, there has been extraordinary growth in the use of technological equipment in the home environment, facilitated by the increasing availability of low-cost appliances and increasing household income levels (Bittman et al., 2004; Selvanathan et al., 2021;

M. Jayasinghe, *Poverty, Food Consumption, and Economic Development*,
https://doi.org/10.1007/978-981-16-8743-3_5

Vanek, 1978). This phenomenon has also been seen in developing nations, including households in rural areas.

Beyond directly improving the welfare of the consumer, this study empirically considers the possibility that these new technologies have impacted the distribution of poverty across large and small households. In particular, this study argues that the use of modern technologies such as microwave ovens, refrigerators, etc. has enabled large households to achieve significantly greater economies of scale in consumption, relative to small households. As a consequence, the adoption of new technologies, such as access to electricity, enables households to realise economies of scale in consumption and improve their living standards. There has been a rapid growth in the emergence of new goods and services in the twentieth century and their diffusion throughout the developing world (Bils & Klenow, 2001; Popkin, 1999). Therefore, it is worth considering how this has affected the ability of large households to achieve lower per capita costs in maintaining a given living standard. To date, much of the literature considering the welfare impact of electronic goods has mainly focused on how new technologies have delivered higher welfare by expanding the choice set faced by households (Hausman & Leibtag, 2007). Less has been done to study the impact of new technologies on household economies of scale.

An analysis of the link between electricity access, domestic technology adoption and consumption economies of scale, and the implications on poverty measurements is important for three reasons. Firstly, the incorporation of economies of scale arising from the consumption of household goods appears to have a significant impact on the measurement of household poverty and welfare (Buhmann et al., 1988; Coulteret al., 1992; Drèze & Srinivasan, 1997; Lanjouw & Ravallion, 1995; Meenakshi & Ray, 2002). Conventional poverty measurements have often been criticised for overestimating poverty levels due to ignoring the possibilities of consumption economies of scale. Therefore, the negligence in identifying the impact of technology adoption in the domestic sphere on consumption economies of scale may lead to a further overestimation of poverty levels. Secondly, an increasing number of households around the world, particularly in developing countries, are now using more technological equipment to ease the burden of domestic tasks. Therefore, it is important to analyse how this increasing trend in domestic technology adoption has positively impacted their poverty levels. Thirdly, if there is adequate evidence to show that the consumption economies of scale deriving from technology adoption aids in curbing poverty levels, the adoption of appropriate technology in the domestic sphere can be promoted as a poverty reduction strategy.

The objective of this chapter is twofold: (1) to theoretically and empirically investigate the effects of technology adoption in the domestic sphere on food consumption scale economies using Engel equivalence scales; and (2) to re-estimate the poverty rates for Sri Lanka to quantify the consumption economies of scale effect on existing poverty levels in the country. This chapter is structured as follows. Section 5.2 establishes the link between electricity access, domestic technology adoption and economies of scale in food consumption. Section 5.3 presents a utility-maximising model of households that highlights how the adoption of domestic technology enhances economies of scale in food consumption. Section 5.4 presents the empirical methodology adopted. Section 5.5 presents the results and discussion; Sect. 5.6 provides concluding remarks and policy implications.

5.2 Review of Literature

People have greatly benefitted from the adoption of domestic technology in the household sphere, such as the use of electrified technology in performing domestic activities (Bittman et al., 2004; Vanek, 1978). These technologies include appliances such as rice cookers, stovetops, convection ovens, microwave ovens, deep freezers, and refrigerators used in food preparation, processing, and preservation, and washing machines and clothes dryers used in the laundry. Televisions and computers used for educational and leisure activities are also considered domestic technology. Over time, there has been an astronomical growth in the use of technological equipment in the home environment, facilitated by the increasing availability of low-cost appliances and increasing household income levels. The growing access to electricity, particularly in developing nations, has also enabled households to use electrical appliances at home, especially the households in rural areas (Wickramasinghe, 2011).

Domestic technology adoption has, in many ways, eased everyday household tasks, particularly by reducing the amount of labour required in the accomplishment of these tasks (Vanek, 1978). For example, cooking food and related tasks, such as gathering firewood, consume a significant amount of female's time, particularly in the rural areas of developing countries. For example, Laxmi et al. (2003) show that in India, women, on average, spend about 15.6 h on firewood collection per month. Most of these tasks that used to be done manually are now performed using machines, saving time that can be spent on other tasks such as an income-earning activity or even on leisure (Hossain, 2010; Wickramasinghe, 2011). It is well acknowledged and proven that domestic technology adoption has improved the welfare of households, in many ways, including enhanced health and educational attainments, gender equality, environmental sustainability, and economic growth, particularly in developing countries (Foell et al., 2011; Wickramasinghe, 2011).

As discussed in previous chapters, there are three main sources of consumption economies of scale; (1) consumption of public goods, (2) increasing returns in home production and (3) bulk purchases (Nelson, 1988). The existing literature provides basic intuition on a number of possible ways in which technological adoption in the domestic sphere, facilitated by domestic electricity access, may enhance greater household economies of scale in relation to the aforementioned three sources. Some studies show that food preparation, preservation, and processing time and costs have seen a considerable reduction over the past several decades, particularly due to the advancements in domestic technology and access to efficient energy sources (Popkin, 1999). For example, access to refrigerators and deep freezers has allowed households to benefit from greater economies of scale in consumption in meal production at home (Bittman et al., 2004). Several studies in the literature provide some evidence that bulk purchases of food have been on the rise in recent years due to the increased availability of appliances (such as refrigerators and freezers) to store perishable food items and due to increased access to transport (own automobiles and public transport services) (Pena & Ruiz-Castillo, 1998; Reardon et al., 2003; Tuttle, 2011). Cooking meals at home has become efficient with the use of electric stoves, microwave ovens and utensils such as grinders and blenders. The use of microwaves, for example, has enabled households to reduce food waste by facilitating an easy way to re-heat excess food for later consumption. These developments may have enabled households to make more efficient use of scarce time and to perform meal preparation related tasks with less toil. As a result, these households may be able to reap economies of scale in food consumption through increasing returns to scale in home production (Vernon, 2005). Nevertheless, a comprehensive analysis of this matter with empirical evidence is still lacking, particularly in the context of developing countries. This study intends to fill this gap in the existing literature by theoretically and empirically investigating the effects on economies of scale in food consumption of technology adoption in the domestic sphere.

5.3 Theoretical Model

The theoretical model used in this chapter is adapted from Jayasinghe et al. (2018). A household comprises n identical individuals who derive utility by consuming two goods, namely, food (quantity x_1) and non-food items (quantity x_2). The household preferences are given by a Cobb–Douglas utility function as

$$u(x_1, x_2) = x_1^a x_2^b \tag{5.1}$$

The power coefficient a denotes the demand elasticity of food and b denotes the demand elasticity of non-food items. Households maximise the utility function (5.1) subject to the budget constraint,

$$p_1 x_1 + p_2 x_2 = I \tag{5.2}$$

where p_1 is price of food, p_2 is price of non-food items, and I is household income.

As economies of scale in food consumption is the main focus of this chapter, we solely concentrate on the demand model for food. The household food consumption is determined by the food price, income, and preferences of the household unit. The optimal consumption of food x_1, based on the utility maximisation problem subject to the budget constraint yields

$$x_1 = \frac{a}{a+b}\frac{I}{p_1} \tag{5.3}$$

This optimal consumption of x_1 can also be referred to as the "unadjusted" household food consumption, as it is not adjusted for possible economies of scale. However, in the presence of economies of scale, the "actual" or "adjusted" household consumption is different from the optimal consumption level given by Eq. (5.3). What follows illustrates how the optimal consumption can be adjusted to incorporate economies of scale in food consumption.

Economies of scale may occur as a result of the following two factors:

(i) Household size (n): This refers to the ability of large households to enjoy consumption economies of scale from bulk purchases, quantity discounts, and substitution of home-produced meals with expensive food-away-from-home (Nelson, 1988);

(ii) Degree of technology adoption in the domestic sphere (τ): This refers to the ability of households to achieve economies of scale by adopting technology to assist in food consumption related aspects. For example, having access to electricity allows them to purchase perishable food items in bulk and to store them for a long time. Furthermore, the increasing returns to scale in the home production of food, which is also a source of economies of scale, may depend on the access to domestic technology such as household utensils and appliances that save the time and effort of the people who are engaged in food preparation activities. For example, if there are electric stoves at home, the household saves time spent not only on cooking using traditional fires but also on not having to collect firewood for cooking purposes. The enhanced time efficiency in food production at home may therefore encourage households to eat more home-prepared meals. All these factors are composited in this model which investigates the effects of technology adoption.

Next, the possibilities of economies of scale in food consumption arising due to household size and technology are introduced to the model and household optimal food consumption is presented in terms of "actual" or "adjusted" household food consumption. Such economies of scale effects are represented by the ϕ parameter. The economies of scale arising from household size n and the degree of technology adoption in the domestic sphere τ are modelled as:

$$\phi(n, \tau) = n^{1-\sigma_1}(1+\tau)^{1-\alpha_1} \tag{5.4}$$

where n is household size and $n^{1-\sigma_1}$ is the economies of scale effect of household size, τ is the degree of technology adoption, and $\tau^{1-\alpha_1}$ is the economies of scale effect of the technology adoption in the domestic sphere. Here σ_1 is the scale elasticity of food with respect to household size, while α_1 is the scale elasticity of food induced by the use of domestic technology to aid the consumption of food. Both these scale elasticities are assumed to range between 0 and 1 (i.e. $0 \leq \sigma_i \leq 1$ and $0 \leq \alpha_i \leq 1$). In the context of modelling of the degree of technology adoption, τ is a parameter that can vary from 0 to 1, where 1 indicates a fully automated food preparation process.

Consequently, the economies of scale parameter $\phi(n, \tau)$ can range from $n(1+\tau)$ to 2; i.e. $n(1+\tau) \leq \phi_1(n, \tau) \leq 2)$, where $\phi(n, \tau) = n(1+\tau)$ implies no economies of scale and $\phi(n, \tau) = 2$ implies full consumption economies of scale, arising from both household size and technology. For example, if the good is a pure public good that provides full economies of scale (i.e. $\sigma_1 = 1$), and the consumption of that good involves a fully automated processes with $\tau = 1$ and scale elasticity $\alpha_1 = 1$, then the scale parameter is $\phi(n, \tau) = 2$. A desktop computer, which can be shared among the household members, and which requires technology to assist deriving utility, represents a conceivably good example for this type of public good.[1]

On the other hand, if the good is a pure private good, no economies of scale generate from household size n (i.e. $\sigma_i = 0$) and no economies of scale effect is derived from technology (i.e. $\alpha_i = 0$), as the degree of technological adoption is 0 (i.e. $\tau = 0$) and so Eq. (5.4) simplifies to $\phi(n, \tau) = n$. This implies that the good is a private good that cannot be shared; hence the good must be fully replicated if all the members in the household are to enjoy the good to the same degree as a single person household; also no technology is involved in consumption of this good. There is another extreme situation where $\sigma_1 = 0$, yet $\tau > 0$, and hence the scale parameter $\phi(n, \tau)$ can be affected. If the good is a pure private good and no economies of scale effect is derived from household size n (i.e. $\sigma_1 = 0$), and no economies of scale effect emerges from use of technology (i.e. $\alpha_1 = 0$), although the household has full access to technology (i.e. $\tau = 1$), then Eq. (5.4) simplifies to $\phi(n, \tau) = n(1+\tau)$. That is although a household has access to technology, the consumption of some goods does not require any technological input. A sightseeing ticket or a ticket for a sports event that cannot be shared, and that does not require any technological input in deriving utility, represents an example for this type of pure private good.

Based on this discussion, the adjusted/actual household consumption of food x_1^* is given as

$$x_1^* = \frac{x_1}{\phi_1(n, \tau)} \tag{5.5}$$

Substituting Eqs. (5.3) and (5.4) into the above Eq. (5.5) and rearranging gives

$$x_1^* = \frac{a}{a+b} * \frac{I}{p_1} * \frac{1}{n^{1-\sigma_1} * (1+\tau)^{1-\alpha_1}} \tag{5.6}$$

The notation x_1^* denotes the adjusted household demand for food. Equation (5.6) can be presented in per capita terms to show the adjusted per capita demand for food as

$$\frac{x_1^*}{n} = \left(\frac{a}{a+b} * \frac{I}{p_1} * \frac{1}{n^{1-\sigma_1} * (1+\tau)^{1-\alpha_1}}\right)\frac{1}{n} \tag{5.7}$$

[1] This example is drawn from non-food category as food that has a higher private component is unlikely to provide full economies of scale in consumption.

When taking the log of both sides, the log of adjusted per capita food consumption is

$$\ln\left(\frac{x_1^*}{n}\right) = \ln(aI) - \ln(a+b) - \ln(p_1) - [(2-\sigma_1)\ln n] - [(1-\alpha_1)\ln(1+\tau)] \tag{5.8}$$

Since $0 \leq \sigma_1 \leq 1$, the fourth term of this equation suggests that per capita food expenditure holds a negative relationship with (log of) household size. Because there are economies of scale in food consumption, the per capita spending on food will decline as household size grows. As $0 \leq \alpha_1 \leq 1$, the negative coefficient in the fifth term suggests that the per capita food expenditure is inversely related to the (log of) degree of technology adoption in the domestic sphere. This is because household food consumption and food production assisted by domestic technology enable households to realise greater economies of scale in consumption. As a result, per capita food consumption declines with the adoption of more technology.

Equation (5.8) is then estimated empirically, in order to verify this theoretical proposition.[2] The (log of) PCE on food is taken as the dependent variable with (log of) household size n and (log of) technology share τ as independent variables. If the estimated coefficients of these two parameters are found to have a negative relationship with per capita food expenditure, this would indicate that household size, as well as adoption of domestic technology to assist food consumption, enhances economies of scale in food consumption. If domestic technology was completely absent, households would still enjoy economies of scale in consumption generating from household size. However, those households which adopt more domestic technology in relation to food consumption can achieve higher economies of scale in food consumption, thereby lowering their per capita food consumption expenditure more than those who do not use domestic technology in this way. In this study, access to electricity is considered as a proxy to indicate domestic technology adoption, while expenditure share on electricity is used as a proxy variable for the degree of technology adoption τ. Firth et al. (2008) showed that the overall increase in domestic electricity consumption is linked with increased household consumption of electrical appliances such as refrigerators, freezers and televisions. In the 2016 HIES dataset used in this study, the expenditure share on electricity and the number of household electrical appliances—those that are directly related to food consumption such as refrigerators, freezers, electric cooker and stoves, grinders and mixers—revealed a positive correlation; hence the expenditure share on electricity appears to be a reasonable proxy for household expenditure share on domestic technology.

---

[2] The empirical estimation adopts a modified version of Eq. (5.8) as we do not have access to food price data and the economies of scale elasticity of food and technology adoption is also unknown. See Sect. 5.4 for more details related to the modified equation used.

Table 5.1 Summary statistics

(1)	National (2)	Urban (3)	Rural (4)	Estate (5)
Percentage of households with access to electricity	95.8	99.2	96.2	92.8
Electricity expenditure as a percentage of total household expenditure	0.02	0.03	0.02	0.01

Source Author's compilation based on Department of Census and Statistics (2017a)

5.4 Data: Household Income and Expenditure Survey HIES (2016)

The empirical analysis of this chapter is based on the 2016 HIES data, discussed in detail in Chap. 3. As the main focus of the chapter is household access to electricity, Table 5.1 provides some information on household electricity consumption behaviour. As can be seen about 96% of households in the sample have access to electricity. At the sectoral level, the urban sector has highest percentage of households (as a percentage of total households in the urban sector), followed by the rural and estate sectors (as a percentage of total households in the respective sectors). Overall, households spend about 1–3% of their total household budget on electricity in Sri Lanka.

5.5 Empirical Methodology

This section details the methodology used for the empirical analysis of this chapter.

5.5.1 Effects of Technology on Food Consumption Economies of Scale

In this section, an empirical methodology is developed to test the theoretical model derived in Sect. 5.3, under a multiple regression framework. The log of per capita food expenditure $\ln\left(F_{pc}\right)$ is used as the dependent variable with a set of independent variables, including log of household size $\ln(n)$ and log of expenditure share on electricity $\ln(EE_{share})$. As discussed in Sect. 5.3, the $\ln(EE_{share})$, a proxy variable for the expenditure share on technology, is expected to demonstrate a negative relationship with $\ln\left(F_{pc}\right)$.[3] This is because of the consumption economies of scale arising from the adoption of domestic technology to facilitate food preservation and

[3] Note that a constant "1" was added to the electricity expenditure share before log transforming, i.e. $\ln(1 + EE_{share})$ was used in the actual estimation. This is to circumvent the problem of zero values being log transformed as missing or undefined values (Hamilton, 2008).

preparation activities. The variable $\ln(n)$ is also expected to possess a negative relationship with $\ln(F_{pc})$ due to household size induced consumption economies of scale in food consumption. The other independent variables used in the model include log of PCE, $\ln(E_{pc})$, and a vector of demographic variables $\boldsymbol{K}$ which include the gender of the head of the household, as well as two location dummy variables for the rural and estate sectors. ε_i is the random error term. The resulting regression specification is as follows:

$$\ln(F_{pc}) = \beta_0 + \beta_1 \ln(n) + \beta_2 \ln(EE_{share}) + \beta_3 \ln(E_{pc}) + \boldsymbol{\delta}_i \boldsymbol{K} + \varepsilon \tag{5.9}$$

There is an issue of endogeneity with the model given by Eq. (5.9). This is because the food expenditure is a part of total expenditure (which includes both food and non-food expenditure), the errors in per capita food expenditure F_{pc}, and PCE or (E_{pc}) may inevitably be correlated. To address possible endogeneity issues arising from the inclusion of PCE, the log of PCI was used as an IV for PCE. This is because income is highly correlated with household expenditure; however, as it is measured independently of expenditure it serves as a good instrumental variable for expenditure (Deaton & Paxson, 1998; Lewbel, 1989). Section 5.6.1 presents the results of this empirical estimation using both Ordinary Least Squares (OLS) method (Model 1) and IV regression method (Model 2).

5.5.2 *Effects of Household Electrification on Food Consumption Economies of Scale*

This section details an alternative approach to identify the effect of household electrification on food consumption economies of scale. That is the Engel equivalence scales approach discussed in Chap. 3 and an adaptation of Deaton and Muellbauer (1980), Jayasinghe et al. (2016) and Lelli (2005). For this purpose, the Engel curve for food was estimated using equations at both the national and the sectoral level.

$$w_f = \beta_0 + \beta_1 \ln\frac{x}{n} + \beta_2 \left(\ln\frac{x}{n}\right)^2 + \gamma_1 n_a + \gamma_2 n_c + \gamma_3 n_a n_c + \varepsilon \tag{5.10}$$

where w_f refers to budget share on food, $\ln\frac{x}{n}$ is the logarithm of PCE, $\left(\ln\frac{x}{n}\right)^2$ is the square of the logarithm of PCE, n_a and n_c is the number of adults and children in the household, respectively, and $n_a n_c$ is the interaction term between the number of adults and children. Equation (5.10) was estimated separately for households with and without access to electricity. Households reporting positive expenditure for electricity were considered as households with access to electricity. The interaction term, $n_a n_c$, is expected to capture the joint effect of adults and children on food budget share. The joint effect on the expenditure share on food in households that have both adults and children may be different to the individual effect of adults and children on the expenditure share on food. These Engel curve estimates were then used to estimate income-dependent equivalence scales as discussed in Sect. 2.3.3. The equivalence scales were calculated for households with and without access to electricity, using the regression results to identifying the effects of household electrification on food consumption economies of scale.

These equivalence scales were compared at both the national and sectoral level to validate the empirical results and investigate the specific variations of results, especially at the sectoral level.

5.5.3 *Impact on Poverty Measurements*

To investigate the effects of food consumption economies of scale associated with electrified households on the measurement of poverty, the percentage of poor households adjusted for consumption economies of scale was estimated. The aim here was to identify how many households normally assessed as living below the official poverty line may actually be placed above the poverty line when food consumption economies of scale derived from the use of electricity are taken into account. To this end, the number of poor households is first estimated using PCE. The OPL for Sri Lanka in 2016 is SLR.4,166 real total expenditure per person (Department of Census and Statistics, Sri Lanka, 2017b).

To adjust the PCE of these households for consumption economies of scale arising from the use of electricity to facilitate food consumption, the expenditure per equivalent adult (EPEA) for food was calculated for households who have access to electricity and who do not have access to electricity, using the equivalence scales derived from respective regression estimates discussed in Sect. 5.5.2 (taking one adult household as the reference household). The EPEA was estimated by dividing the household expenditure by the equivalence scales of respective households, instead of by household size (Jayasinghe et al., 2019). For the sake of simplicity, it was assumed that economies of scale in consumption were absent in other expenditure categories, and hence the use of PCE for expenditure categories other than the expenditure on food. Then we investigate the change in the percentage of poor households under three scenarios; (1) PCE is used, (2) EPEA with electricity is used, and (3) EPEA without electricity is used.

5.6 Results and Discussion

This section presents estimation results that explain the impact of domestic technology adoption and household access to electricity on food consumption economies of scale and implications on poverty measurements.

5.6.1 *Effects of Technology on Food Consumption Economies of Scale*

Table 5.2 presents estimation results for Eq. (5.9) estimated using the OLS (Model 1) and IV estimation methods (Model 2). Considering the OLS regression estimates, the household size elasticity of per capita food β_1 is negative, as expected in the theoretical model, implying that household size is negatively related with the per capita expenditure on food. This reflects the food consumption economies of scale with respect to household size and as a result, large households can achieve the same standard of living (defined in terms of food consumption, in this case) with lower per capita food expenditure. As suggested in the theoretical model developed in Sect. 5.3, the (log of) the electricity expenditure share β_2 also takes a negative value, suggesting that a 10% increase in expenditure share on electricity is associated with a 2.8% reduction in PCE on food. As the theoretical model suggested, this reduction in per capita food expenditure is due to consumption economies of scale arising from technological adoption in the domestic sphere to facilitate food consumption related activities.

As noted in Sect. 5.5.1, an IV regression model is also estimated as a robustness test and to address the potential endogeneity issue associated with the OLS estimation. Here, the log of PCI was used as an instrumental variable for log of PCE and the IV estimation results are given under Model 2 in Table 5.2. The Wu-Hausman test, with $F(1{,}19{,}944) = 10.2287$ and ($p = 0.0006$), indicates that the null hypothesis that PCE is exogenous is rejected, indicating that PCE is endogenous. However, the sign of the household size and electricity expenditure share variables in the IV regression are consistent with the theoretical model and those of the OLS estimates.

Table 5.2 Effects of technology adoption on food consumption economies of scale

	Model 1 (OLS)		Model 2 (IV regression)	
	Coefficient	*p*-value	Coefficient	*p*-value
Constant (β_0)	2.748*	0.042	3.295*	0.064
Log of household size(β_1)	−0.196*	0.003	−0.201*	0.002
Log of electricity expenditure share (β_2)	−0.287*	0.000	−0.211*	0.000
Log of PCE (β_3)	0.521*	0.001	0.497*	0.002
Vector of demographic variables ($\boldsymbol{\delta}_i$)				
Gender of the head of the household	−0.019*	0.000	−0.021*	0.002
Rural dummy	−0.041*	0.000	−0.038*	0.000
Estate dummy	−0.061*	0.000	−0.053*	0.000
R squared	0.6462		0.6335	
Number of observations	19,952		19,952	

Note Dependent variable is the per capita food expenditure. * $p < 0.05$
Source Author's compilation based on Department of Census and Statistics (2017a)

5.6.2 *Effects of Household Electrification on Food Consumption Economies of Scale*

Table 5.3 presents the parameter estimates of the Engel curve for food corresponding to Eq. (5.10) at the national and sectoral levels. These results are derived from models that have been corrected for heteroskedasticity. The adjusted R-squared values of between 34 and 61% indicate that the estimated Engel curves for food fit reasonably well to the cross-sectional Sri Lankan data. Part A presents estimation results for

Table 5.3 Food Engel curve estimates for households with and without electricity

Variable	National	Urban	Rural	Estate
Part A: *households with electricity*				
Constant (β_0)	0.984* (0.000)	1.917* (0.001)	0.837* (0.001)	−1.251 (0.109)
Log of PCE (β_1)	−0.038 (0.267)	−0.145* (0.001)	−0.013 (0.955)	0.340* (0.000)
Square of Log of PCE (β_2)	−0.008* (0.000)	−0.001 (0.675)	−0.010* (0.000)	−0.035* (0.000)
Number of adults (γ_1)	−0.028* (0.000)	−0.021* (0.000)	−0.031* (0.000)	−0.029* (0.000)
Number of children (γ_2)	−0.041* (0.000)	−0.024* (0.000)	−0.048* (0.000)	0.030* (0.000)
Interaction term—number of children and adults (γ_3)	0.007* (0.000)	−0.005* (0.000)	0.007* (0.000)	0.004* (0.000)
R-squared	0.611	0.564	0.587	0.521
Part B: *households without electricity*				
Constant (β_0)	−0.986* (0.015)	1.056* (0.048)	−2.342* (0.000)	−1.987* (0.049)
Log of PCE (β_1)	−0.020* (0.000)	−0.088 (0.580)	0.432* (0.000)	0.675* (0.001)
Square of Log of PCE (β_2)	−0.003* (0.000)	−0.013 (0.675)	−0.053* (0.000)	−0.065* (0.000)
Number of adults (γ_1)	−0.032* (0.000)	−0.026* (0.000)	−0.036* (0.000)	−0.027* (0.001)
Number of children (γ_2)	−0.041* (0.000)	−0.028* (0.039)	−0.048* (0.000)	−0.034* (0.000)
Interaction term—number of children and adults (γ_3)	0.005* (0.000)	0.002 (0.376)	0.014* (0.000)	0.007* (0.000)
R-squared	0.432	0.415	0.342	0.354

Note Dependent variable is the expenditure share on food for household with (Part A) and without electricity (Part B) at the national and sectoral level. p-values are in parenthesis. *$p < 0.05$
Source Author's compilation based on Department of Census and Statistics (2017a)

households with access to electricity while Part B presents results for households who do not have access to electricity.

When considering the coefficient estimates of the $\ln\frac{x}{n}$ and $(\ln\frac{x}{n})^2$ variables, most of the coefficients are statistically significant at the 5 or 10% levels. Engel's law suggests that the food share declines as household income increases. Accordingly, it was expected that the β_1 and β_2 coefficients would be negative. For example, at the national level, a 10% increase in PCE increases the food share by 3.8% for households who have access to electricity. The negative signs of the β_2 coefficients coupled with negative β_1 coefficients indicate that the food share decreases at a decreasing rate with increasing household income levels.

With respect to family composition, the negative sign of the γ_1 and γ_2 coefficients implies that when the household size increases by one adult (or child), the expenditure share on food declines. For example, at the national level, an increase in household size by one adult reduces the food share by 0.032, with other things held constant. This observed inverse relationship between the number of adults (n_a) and the number of children (n_c) variables and the food budget share is what is anticipated based on the literature. This implies that due to economies of scale in food consumption, an additional adult or child adds less than double to the household expenditure on food. When considering the size of the coefficients of both variables, it is evident that the coefficients of γ_2 are consistently larger than those of γ_1. This implies that the change in food expenditure share for an additional child is greater than that of an additional adult. This observation is not surprising, due to the fact that the food consumption needs of adults are greater than those of children. Furthermore, the results suggest that the joint impact of the interaction term on food expenditure share is negative. At the national level, for example, if it is assumed that the number of adults is 1 and then there is an increase in number of children by 1 child, this reduces the food share by 0.020 ($-0.028 + 0.007$) *ceteris paribus.* With the presence of economies of scale in food consumption, the joint negative effect of the interaction term on the food expenditure share is in line with expectations.

Table 5.4 records the marginal equivalence scales estimated using the national and sectoral level Engle curve estimations given in Table 5.4. Part A presents the marginal equivalence scales for households with access to electricity while Part B presents those for households without access to electricity. The marginal equivalence scale refers to the change in the cost of remaining at the same welfare level (measured in terms of food consumption) when household size increases by one adult. In this case, it is assumed that all of the adults in the household have identical tastes, irrespective of their gender and age.

Equivalence scales are estimated using the mean household expenditure at the national and the respective sector levels. In Part A, the marginal equivalence scale of 0.38 for the 1,0–2,0 household size at the national level implies that when a household size increases from single adult (1,0) to two-adult (2,0), the two-adult household with access to electricity is required to spend 38% more on food to remain in the same level of welfare (measured in terms of the budget share on food) of a one-adult household.

The results also reveal that the households with access to electricity show lower equivalence scales than those of the households without access to electricity. For

Table 5.4 Marginal equivalence scales of the households with and without electricity

Household size	National	Urban	Rural	Estate
Part A: *marginal equivalence scales of the households with electricity*				
1,0–2,0	0.38 (0.000)	0.40 (0.000)	0.38 (0.001)	0.41 (0.000)
2,0–3,0	0.22 (0.001)	0.26 (0.002)	0.21 (0.001)	0.26 (0.001)
3,0–4,0	0.11 (0.001)	0.15 (0.002)	0.10 (0.000)	0.15 (0.000)
Part B: *marginal equivalence scales of the households without electricity*				
1,0–2,0	0.39 (0.002)	0.42 (0.001)	0.39 (0.002)	0.42 (0.002)
2,0–3,0	0.24 (0.001)	0.30 (0.000)	0.23 (0.001)	0.27 (0.003)
3,0–4,0	0.13 (0.000)	0.20 (0.001)	0.12 (0.001)	0.16 (0.001)

Note Bootstrapped standard errors are given in parenthesis
Source Author's compilation based on Department of Census and Statistics (2017a)

example, when household size increases from (2,0) to (3,0), an urban household with access to electricity incurs 26% more expenditure on food to remain in the same welfare level as of (2,0), while a household without access to electricity spends 30% more on food to remain in the same welfare level as a household of (2,0). This result implies that households with access to electricity enjoy greater food consumption economies of scale through increased use of electrical appliances to facilitate food preparation and hence spend less on food for an additional adult to remain in the same welfare level compared to a household without access to electricity. This observation is consistent at the national and sector level across all household types. The standard errors, given in parenthesis, estimated using the bootstrapping method with 1000 replications, indicate that the equivalence scales are statistically different from each other.

5.6.3 Impact on Poverty Measurement

As discussed in Chap. 2, one of the major implications of economies of scale in consumption is their effects on poverty measurements. This section presents the estimation results on the effect of food consumption economies of scale on the percentage of poor households, in particular for households with and without access to electricity.

Two main observations emerge from the results on poverty estimates as presented in Table 5.5. Firstly, the results reveal that, overall, when the necessary adjustments for economies of scale in food consumption in households (with and without access

Table 5.5 Impact of food consumption scale economies on poverty status, by survey year

Percentage of poor households (1)	National (2) (%)	Urban (3) (%)	Rural (4) (%)	Estate (5) (%)
Part A: *households with access to electricity*				
PCE	6.5	2.5	6.8	13.1
EPEA	1.1	0.4	1.2	2.0
Percentage decline	*5.4*	*2.1*	*5.6*	*10.9*
Part B: *households without access to electricity*				
PCE	8.1	3.7	7.9	15.7
EPEA	3.9	1.9	3.7	7.6
Percentage decline	*4.2*	*1.8*	*4.2*	*8.1*

Source Author's compilation based on Department of Census and Statistics (2017a)

to electricity) are allowed for, that is when EPEA is used, there is a considerable reduction in the percentage of households rated as poor at the national and the sectoral level. At the national level, for example, for households with access to electricity, the reduction in the poverty head-count rate is about 5.4% when EPEA is used. A similar pattern can also be seen at the sectoral level. This observation is in line with the findings in the literature that often poverty measures based on the PCE overestimate the poverty levels of large households as PCE method does not take possible CES into account (Jayasinghe, 2019; Jayasinghe & Smith, 2021; Jayasinghe et al., 2017; Lanjouw & Ravallion, 1995). Secondly, the difference between the percentage of poor households under PCE and EPEA (reported in Table 5.5 as the percentage decline) is higher for households with access to electricity than those without access to electricity at the national level and the sectoral level. For example, at the national level, the decline in the poverty head-count rate for households with access to electricity is 5.4%, while for households without access to electricity it is 4.2%. This observation is not surprising given that households with access to electricity materialise more food consumption economies of scale and hence these households show greater sensitivity for poverty measurements that take possible food consumption economies of scale into account.

5.7 Concluding Remarks and Policy Implications

This chapter examined, both theoretically and empirically, the role of domestic technology adoption, facilitated by electricity access in enhancing food consumption economies of scale. The empirical results revealed that technological adoption in the domestic sphere enables households to harvest greater food consumption economies scale and thereby stretch their food dollar. This is because the adoption of domestic technology together with electrification appears to reduce their food preparation, processing, and preservation cost. Furthermore, the reduction in time required to perform food preparation related activities, with the use of domestic technology, may be encouraging time scarce households to prepare food at home rather than having meals prepared outside. This may enable these households to further reap the benefits of consumption economies of scale through increasing returns to scale in home production. Confirming the validity of the hypothesis that household electrification enables greater economies of scale in food consumption, equivalence scales, estimated separately for households with and without electricity, revealed that households with access to electricity demonstrate lower equivalence scales and thus higher economies of scale in food consumption than households without access to electricity. The results on poverty estimates revealed that when appropriate adjustments are made to allow for the possibilities of economies of scale in food consumption, the percentage of poor households appears to decline significantly.

The effects of technology adoption have commonly been examined in the context of economies of scale in production. To this end, the analysis carried out in this chapter sheds some new light on the welfare implications of household electrification and domestic technology adoption. The findings of this study provide some important policy insights on the importance of encouraging domestic technology adoption. One approach to encourage domestic technology adoption is the further extension of subsidies that allow for a greater level of access to electricity in disadvantaged areas within the rural and estate sectors and provision of basic household electrical equipment at subsidised prices for very poor households. This could act to improve the standard of living of households in a well-targeted manner.

References

Annemarije, V. D. (2008). *The power to produce: The role of energy in poverty reduction through small scale enterprises in the Indian Himalayas*. University of Twente.

Awaworyi Churchill, S., Smyth, R., & Farrell, L. (2020). Fuel poverty and subjective wellbeing. *Energy Economics*, *86*(C), 104650.

Barnes, D. F., Khandker, S. R., Samad, H., & Minh, N. H. (2009). Welfare impacts of rural electrification: evidence from Vietnam (A. S. a. A. E. A. Program, Trans.) *Policy Research Working Paper* (Vol. IE 38). The World Bank

Bils, M., & Klenow, P. J. (2001). The acceleration in variety growth. *The American Economic Review*, *91*(2), 274–280. https://doi.org/10.1257/aer.91.2.274

Bittman, M., Rice, J. M., & Wajcman, J. (2004). Appliances and their impact: The ownership of domestic technology and time spent on household work. *The British Journal of Sociology*, *55*, 401–423.

Buhmann, B., Rainwater, L., Schmaus, G., & Smeeding, T. M. (1988). Equivalence scales, well-being, inequality, and poverty: Sensitivity estimates across ten countries using the Luxembourg Income Study (LIS) Database. *Review of Income and Wealth*, *34*(2), 115–142.

Cook, C. C., Duncan, T., Jitsuchon, S., Sharma, A., & Guobao, W. (2005). *Assessing the impact of transport and energy infrastructure on poverty reduction*. Asian Development Bank.

Coulter, F. A. E., Cowell, F. A., & Jenkins, S. P. (1992). Differences in needs and assessment of income distributions. *Bulletin of Economic Research*, *44*(2), 77–124. https://doi.org/10.1111/j.1467-8586.1992.tb00538.x

De Vries, J. (2008). *The industrious revolution: Consumer behavior and the household economy, 1650 to the present*. Cambridge University Press.

Deaton, A., & Muellbauer, J. (1980). *Economics and consumer behavior*. Cambridge University Press.

Deaton, A., & Paxson, C. (1998). Economies of scale, household size, and the demand for food. *The Journal of Political Economy, 106*, 897–930.

Department of Census and Statistics (DCS). (2017a). *Household income and expenditure survey—2016*. Department of Census and Statistics.

Department of Census and Statistics (DCS). (2017b). *Official poverty line by district*. Department of Census and Statistics.

Drèze, J., & Srinivasan, P. V. (1997). Widowhood and poverty in rural India: Some Inferences from household survey data. *Journal of Development Economics, 54*(2), 217–234. https://doi.org/10.1016/S0304-3878(97)00041-2

Fan, S., Nyange, D., & Rao, N. (2005). Public investment and poverty reduction in Tanzania: evidence from household survey data', *DSGD Discussion Paper*. International Food Policy Research Institute.

Firth, S., Lomas, K., Wright, A., & Wall, R. (2008). Identifying trends in the use of domestic appliances from household electricity consumption measurements. *Energy & Buildings, 40*(5), 926–936. https://doi.org/10.1016/j.enbuild.2007.07.005

Foell, W., Pachauri, S., Spreng, D., & Zerriffi, H. (2011). Household cooking fuels and technologies in developing economies. *Energy Policy, 39*(12), 7487–7496. https://doi.org/10.1016/j.enpol.2011.08.016

Hall, B. H., & Khan, B. (2003). Adoption of new technology. *NBER Working Paper Series*. National Bureau of Economic Research.

Hamilton, L. C. (2008). *Statistics with Stata: Updated for version 10*. Brooks/Cole: Cengage Learning.

Hausman, J., & Leibtag, E. (2007). Consumer benefits from increased competition in shopping outlets: Measuring the effect of Wal-Mart. *Journal of Applied Econometrics, 22*(7), 1157–1177. https://doi.org/10.1002/jae.994

Hossain, M. M. (2010). Role of technology in consumption and everyday life in rural Bangladesh. *Technology in Society, 32*, 130–136.

Jayasinghe, M. (2019). Dynamic trends in household poverty and inequality in Sri Lanka: Do gender and ethnicity matter? *Journal of the Asia Pacific Economy, 24*(2), 208–223. https://doi.org/10.1080/13547860.2019.1573716

Jayasinghe, M., & Smith, C. (2021). Poverty implications of household headship and food consumption economies of scale: A case study from Sri Lanka. *Social Indicators Research, 155*, 157–185.

Jayasinghe, M., Smith, C., Chai, A., & Ratnasiri, S. (2016). The implications of income dependent equivalence scales for measuring poverty in Sri Lanka. *International Journal of Social Economics, 43*, 1300–1314.

Jayasinghe, M., Chai, A., Ratnasiri, S., & Smith, C. (2017). The power of the vegetable patch: How home-grown food helps large rural households achieve economies of scale and escape poverty. *Food Policy, 73*, 62–74.

Jayasinghe, M., Ratnasiri, S., Smith, C., & Chai, A. (2018). Domestic technology, consumption economies of scale and poverty: Evidence from Sri Lanka. *Applied Economics, 50*, 1777–1789.

Jayasinghe, M., Selvanathan, E. A., & Selvanathan, S. (2019). An intertemporal analysis of expenditure elasticities under three expenditure specifications for Sri Lanka. *Applied Economics, 51*(59), 6376–6392. https://doi.org/10.1080/00036846.2019.1619020

Jayasinghe, M., Selvanathan, E. A., & Selvanathan, S. (2021). Energy poverty in Sri Lanka. *Energy Economics, 101*, 105450.

Kooijman-Van Dijk, A. L. (2012). The role of energy in creating opportunities for income generation in the Indian Himalayas. *Energy Policy, 4*, 529–536.

Lanjouw, P., & Ravallion, M. (1995). Poverty and household size. *The Economic Journal, 105*, 1415–1434.

Laxmi, V., Parikh, J., Karmakar, S., & Dabrase, P. (2003). Household energy, women's hardship and health impacts in rural Rajasthan, India: Need for sustainable energy solutions. *Energy for Sustainable Development, 7*(1), 50–68.

Lelli, S. (2005). Using functionings to estimate equivalence scales. *Review of Income and Wealth, 51*, 255283.

Lewbel, A. (1989). Household equivalence scales and welfare comparisons. *Journal of Public Economics, 39*, 377–391.

Meenakshi, J. V., & Ray, R. (2002). Impact of household size and family composition on poverty in rural India'. *Journal of Policy Modeling, 24*, 539–559.

Mokyr, J. (2000). Why "more work for mother?" knowledge and household behavior 1870–1945. *The Journal of Economic History, 60*, 1–14.

Nelson, J. A. (1988). Household economies of scale in consumption: Theory and evidence. *Econometrica, 56*, 1301–1314.

Pena, D., & Ruiz-Castillo, J. (1998). The estimation of food expenditures from household budget data in the presence of bulk purchase. *Journal of Business & Economic Statistics, 16*, 292.

Popkin, B. M. (1999). Urbanization, lifestyle changes and the nutrition transition. *World Development, 27*, 1905–1916.

Reardon, T., Timmer, C. P., Barrett, C. B., & Berdegué, J. (2003). The rise of supermarkets in Africa, Asia, and Latin America. *American Journal of Agricultural Economics, 85*, 1140–1146.

Selvanathan, S., Selvanathan, S., & Jayasinghe, M. (2021). A new approach to analyse conditional demand: An application to Australian energy consumption. *Energy Economics, 93*, 1–8. [105037]. https://doi.org/10.1016/j.eneco.2020.105037

Tuttle, B. (2011). *Buying in Bulk: Trimming Food Bills Beyond the Warehouse Stores* [Online]. Retrieved March 3, 2015, from http://business.time.com/2011/01/13/buying-in-bulk-trimming-food-bills-beyond-the-warehouse-stores/

Vanek, J. (1978). Household technology and social status: Rising living standards and status and residence differences in housework. *Technology and Culture, 19*, 361–375.

Vernon, V. (2005). *Food expenditure, food preparation time and household economies of scale*. State University of New York.

Wickramasinghe, A. (2011). Energy access and transition to cleaner cooking fuels and technologies in Sri Lanka: Issues and policy limitations. *Energy Policy, 39*, 7567–7574.

Chapter 6
Gender, Economies of Scale and Poverty

6.1 Introduction

The gender of the household head has received significant attention in the poverty analysis (Chant & Campling, 1997; Visaria, 1980). Ensuring gender equality is also one of the 17 United Nations Sustainable Development Goals (SDGs) that countries across the world are intending to achieve by the year 2030. Increased attention towards gender, in general, and gender of the household head, in particular, is vital in poverty analysis because; (1) it is widely acknowledged that female-headed households (FHHs) disproportionately contribute to poverty and that they are believed to be highly disadvantaged compared to male-headed households (MHHs) (Buvinic and Gupta, 1997; Chant, 1997, 2003, 2004) and (2) household headship is also associated with changes in household size and composition, which in turn has significant implications on poverty measurements and welfare. For example, research suggests that FHHs are, in general, smaller than MHHs and that female household headship is associated with a higher dependency ratio (i.e., the presence of a greater number of dependents—both young and old—with fewer adult earners) (Chant, 2004; Chant & Campling, 1997; De Silva, 2005). These matters are becoming even more crucial as there has been an increasing trend in the prevalence of households headed by females worldwide due to various reasons such as gender-selective migration, gender differences in life expectancy, and conflicts and wars (Chant & Campling, 1997; Visaria, 1980).

As discussed throughout this book, household size (and composition) is the key concept related to the consumption economies of scale. This is because large households are able to enjoy higher consumption scale economies than small households, enabling them to enjoy the same standard of living as that of a smaller household, yet with lower PCE. One of the major drawbacks of the traditional approaches of poverty measurements that are based on the per capita methods is that they do not provide avenues to incorporate the effects of the differences in household size and composition associated with the gender of household headship when estimating poverty measures and comparing poverty levels across MHHs and FHHs. The failure to

M. Jayasinghe, *Poverty, Food Consumption, and Economic Development*,
https://doi.org/10.1007/978-981-16-8743-3_6

consider the differences in consumption economies of scale in large and small households often results in an underestimation of the incidence of poverty among the FHHs, which are typically smaller than MHHs (Bongaarts, 2001; Horrell & Krishnan, 2008).

Gender of the household headship may play another significant role in determining the level of consumption economies of scale, particularly those of food in a household. Based on the traditional gender roles, women have been responsible for supplying food for household members and ensuring the nutrition of household members, particularly child nutrition (Kalansooriya & Chandrakumara, 2014; Kennedy & Peters, 1992). In conventional societies, adult women appear to impart traditional knowledge and customs about the preparation and management of household food supply to younger females than to younger males. Such knowledge may act as an expedient to better manage household food supply and thereby reap higher food consumption economies of scale in FHHs, particularly in the absence of a male breadwinner to support the household. These higher economies of scale in food consumption that are likely to be observed in FHHs may make them better off than MHHs, in some respects.

Some research has shown that household income plays a critical role in determining the level of consumption economies of scale (see, for example, Donaldson & Pendakur, 2003; Koulovatianosa et al., 2005; Balli & Tiezzi, 2013; Jayasinghe et al., 2016) where low-income households tend to reap more economies of scale in consumption than high-income households. A considerable amount of research has found that FHHs, on average, have lower income levels than the MHHs (Buvinić & Gupta, 1997; Lampietti & Stalker, 2000). The differences in average incomes by the gender of the household head may, in turn, affect the level of consumption economies of scale materialise by the respective households.

An identification of the variation of consumption economies of scale based on the gender of the household, therefore, enables the precise identification of the poor and the most vulnerable groups in society. This also enables a more accurate comparison of poverty across households with different sizes and compositions, and across other socio-economic and behavioural differences emerging from the differences based on the gender of the household headship.

The objective of this chapter is twofold; (1) to examine how the gender of the head of the household affects economies of scale in food consumption; and (2) to explore how these differences in economies of scale in food consumption, based on the gender of the household head, impact on household poverty status. In particular, this chapter undertakes a further disaggregate analysis by investigating the differences in economies of scale in food consumption by *de-facto* and *de-jure* FHHs. *De-facto* FHHs are those that are headed by women due to the prolonged absence of the adult man and where most of the day-to-day household decisions are made by the female household head. *De-jure* FHHs are households headed by widows, or unmarried, divorced or separated women. Ramnarain (2016) noted that a significant proportion of FHHs are *de-facto* woman heads and these households are more likely to be financially better off than *de-jure* FHHs due to financial support received from the long-distanced partner. As such, *de-facto* and *de-jure* FHHs also can have substantial household size, compositional and income differences (Quisimbing et al., 2001).

The remainder of the chapter is organised as follows. Section 6.2 presents a review of the literature on the link between household headship, poverty and consumption economies of scale. Section 6.3 discusses the data and methodology adopted in this study. Section 6.4 presents the empirical results and a discussion while Sect. 6.5 provides some concluding remarks.

6.2 Review of Literature

There is a significant body of literature concerning female headship and poverty. Research suggests that FHHs, particularly those who are living in rural areas, appear to be in a disadvantageous position in performing household tasks relative to their male counterparts because of lower earnings, limited access to remunerative jobs, and less access to resources such as land, credits, and technology (Buvinić & Gupta, 1997). A considerable number of studies, both in the context of developed and developing countries, have investigated whether households headed by females are poorer than those headed by males. Buvinić and Gupta (1997), for example, reviewed 61 studies about the association between female headship and poverty, covering countries in Africa, Asia, and Latin America and the Caribbean. They showed that 38 of the 61 studies, based on a variety of poverty measurements (such as total or per capita household income and expenditure, income per-equivalent adult, and access to services and ownership of land assets) suggest that FHHs are overrepresented among the poor. Furthermore, Lampietti and Stalker (2000), reviewed 58 World Bank Poverty Assessments that were carried out since 1998 and concluded that 43% of FHHs are poorer than MHHs. Quisumbing et al. (2001) conclude that out of 10 developing countries (Botswana, Cote d'Ivoire, Ethiopia, Ghana, Madagascar, Rwanda, Bangladesh, Indonesia, Nepal, and Honduras), only 2 countries (Ghana and Bangladesh) show that FHHs tend to be poorer than MHHs.

The measurement of poverty based on the gender of the household head appears to vary by the definition of the gender of the household headship. Usually, household surveys record self-declared household headship (as reported by the respondent based on the judgment of household members). As discussed in Buvinić and Gupta (1997), when there is a male adult in the household, he is normally identified as the head of the household. The absence of an adult male in the household, due to various reasons, such as divorce, out-migration, single and never married, single with dependants, etc. forces a female adult in the household to be the head of the household. Although all these households are formally categorised as FHHs, they are a heterogeneous group, where the heterogeneity is largely based on the factors that led them to be household heads. The poverty status of FHHs appears to be sensitive to the underlying reason that led them to have a female head.[1] The lack of an appropriate definition of FHHs

[1] Quisumbing et al. (2001) provide a detailed analysis of two types of female-headed households: *de-jure* (households where the woman is considered the legal and customary head of the household, usually headed by divorced, separated, unmarried or widowed women) and *de-facto* (households

has been highlighted as a major issue relating to the analysis of poverty among FHHs (Buvinić & Gupta, 1997; Haddad et al., 1996; Handa, 1994). Kennedy and Haddad (1994), Krishnan and Horrell (2007), and Van de Walle (2013) showed that in Ghana *de-jure* FHHs, who do not receive income transfers such as child support payment, are the poorest.

In general, female household heads rarely live with a spouse while a majority of male household heads live with their wives, leading to considerable household size and compositional differences across MHHs and FHHs (Bongaarts, 2001; Chant, 2011; Krishnan & Horrell, 2007). Due to household size and compositional differences associated with the gender of the household head (Deaton & Paxson, 1998; Lanjouw & Ravallion, 1995; Quisumbing et al., 2001), the literature shows that FHHs reflect a higher incidence of poverty when the necessary adjustments are made to allow for the impact of possible consumption economies of scale. This is because the poverty levels in smaller households such as FHHs might be underestimated under the traditional approach (Quisumbing et al., 2001). For example, Drèze and Srinivasan (1997) and Van de Walle (2013) showed that although no significant differences in poverty status among the MHHs and FHHs were found under per capita poverty measures, the incidence of poverty among FHHs disproportionately increased once the possibilities for consumption economies of scale were allowed for. This was found to be especially true in the case of single widows and widows living with children, who were more inclined to live in small households.

Other than the size and compositional differences, the economies of scale materialised by households headed by males and females can be different due to various other socio-economic and behavioural factors. The existing literature provides some evidence on the possible reasons for food consumption economies of scale to vary by the gender of the household head.

Some research suggests that income is an important factor that determines the level of consumption economies of scale and that low-income households enjoy higher economies of scale than high-income households. For example, Burch and Matthews (1987) presented a descriptive analysis, which raises the possibility that poor households enjoy greater consumption economies of scale. Beatty (2010) and Griffith et al. (2009) used data from the United Kingdom, and Leibtag and Kaufman (2003) used data from the United States to further confirm this observation that poorer households enjoy greater economies of scale in food consumption. Logan (2011) strengthened the same argument further, showing that equivalence scales for some commodities, including food, increase with income in the United States. Jayasinghe et al. (2016) found that Engel equivalence scales increase with income, suggesting that low-income households materialise more economies of scale in food consumption than high-income households.

Traditionally, women have been responsible for household activities; mainly in performing household food supply related tasks than men (Greenstein, 2000). In particular, women have been playing a vital role in food processing, preparation and

---

where the self-declared male head is absent for a large proportion of the time but may contribute to or even control household finances).

storing and also passing on this knowledge to the next generation, especially to daughters (Van der Lippe & Siegers, 1994). This traditional knowledge includes preserving vegetables, fruits and milk, in times of abundance, for use during scarce time through various indigenous food preservation methods such as drying or fermentation. The use of wild food such as fruits, nuts and leaves to support household food supply is also common in traditional societies. Such food provides economical and nutritious food for the family throughout the year and hence appears to play a major role in food security, particularly in rural societies (Ibnouf, 2012; Kalansooriya & Chandrakumara, 2014). Studies have also found that poorer rural FHHs provide more nutritious food for their children than those headed by men (Kennedy & Peters, 1992). Furthermore, rural women may demonstrate greater indigenous knowledge about food processing and preserving than urban women due to the close social networking system on these practices in the rural areas. This implies that women, particularly those who are in rural areas are more likely to utilise available resources more wisely and thereby improve the nutritional status of the family members (Kalansooriya & Chandrakumara, 2014; Kennedy & Peters, 1992). This, in turn, is expected to deliver greater economies of scale in food consumption among these FHHs. On the other hand, women, particularly those in urban areas, with poor social networking systems on traditional practices, and men in general, who lack such knowledge about traditional food-management skills, may not be able to reap such benefits.

There is a growing body of literature that focuses on household headship and poverty in Sri Lanka, in particular with a reference to a post-war context (see, for example, Amirthalingam & Lakshman, 2013; Aturupane et al., 1997; Gunatilaka and Vithanagama, 2018; Jayathilaka, 2007; Jayasinghe, 2019; Jayasinghe & Smith, 2021; Lakshman & Jayasinghe, 2011; Mahadevan & Jayasinghe, 2020; Ruwanpura & Humphries, 2004; Ruwanpura, 2006, Vasudevan, 2013). These studies provide mixed results in terms of the poverty status of FHHs in Sri Lanka. Aturupane et al. (1997) found that the average monthly household income for poor FHHs is lower than that of male-headed households. Samarasinghe (1993) noted that different groups of Sri Lankan women display nonstandard socio-economic characteristics that are often associated with ethnicity and hence have different poverty levels. Ruwanpura and Humphries (2004) find that FHHs in the Eastern province in Sri Lanka appear to be poorer than the FHHs elsewhere in the country, due to the adverse impact of the war. Jayasinghe (2019) found that while FHHs show a higher prevalence of poverty than their male counterparts in both Sinhala and Tamil households, the Tamil FHHs are more vulnerable to poverty. However, Jayathilaka (2007), using per-capita income to compare poverty, for example, finds that there is no statistically significant difference in poverty levels among MHHs and FHHs at the national level but that considerable regional disparities exist.

The empirical analysis in this chapter addresses two important conceptual challenges for measuring and comparing poverty and welfare of households headed by females and males. Firstly, it addresses the issues pertaining to the definition of the household headship by investigating the poverty levels of MHHs, *de-facto* FHHs and *de-jure* FHHs, separately. Secondly, the poverty figures estimated in this chapter have been adjusted for economies of scale in food consumption, which depends on the household size and compositional differences based on the gender of the household head.

6.3 Data and Methodology

6.3.1 Data: Household Income and Expenditure Survey 2016

The empirical analysis in this chapter is based on the 2016 HIES, which was discussed in detail in Chap. 3. Table 6.1 provides summary statistics for MHHs, *de-facto* and *de-jure* FHHs at the national level and the sectoral level. As can be seen, at the national level, about 9% of the households are *de-facto* FHHs while about 17% of household are *de-jure* FHHs. At the sectoral level, the rural sector has the highest percentage

Table 6.1 Summary statistics by the gender of the household head

(1)	National (2)	Urban (3)	Rural (4)	Estate (5)
Part A: *male-headed households (MHHs)*				
Percentage of households	74.0	11.5	59.3	3.2
Age of head (years)	52.0	51.9	52.2	50.4
Household size	4.2	4.3	4.1	4.5
Dependency ratio	0.4	0.4	0.4	0.4
Total household expenditure (SLR)	49,548	65,973	47,094	36,019
Food expenditure share	0.5	0.4	0.5	0.5
Part B: *de-facto female-headed households (de-facto FHHs)*				
Percentage of households	9.0	1.4	7.1	0.5
Age of head (years)	41.8	43.5	41.5	42.2
Household size	4.5	4.7	4.4	4.8
Dependency ratio	0.6	0.5	0.6	0.6
Total household expenditure (SLR)	47,801	61,360*	46,270*	31,269
Food expenditure share	0.5	0.4	0.4	0.5
Part C: *de-jure female-headed households (de-jure FHHs)*				
Percentage of households	17.0	2.6	13.7	0.7
Age of head (years)	61.0	61.5	61.0	60.9
Household size	3.3	3.6	3.2	3.6
Dependency ratio	0.4	0.4	0.4	0.5
Total household expenditure (SLR)	37,885	50,909	35,827	29,665
Food expenditure share	0.5	0.4	0.5	0.5

Notes Household expenditure data is given in Sri Lankan Rupees (SLR). * Indicates the mean values that are not statistically significantly different from those of MHHs
Source Author's calculations based on Department of Census and Statistics (2017)

of both *de-facto* FHHs (7.1%) and *de-jure* FHHs (13.7%) followed by the urban sector (*de-facto* FHHs, 1.4% and *de-jure* FHHS, 2.6%). The data reveals that, at the national level, *de-jure* female household heads are the oldest (61 years), followed by male household heads (52 years). Along with the differences in household size and composition by gender of the head of the household discussed in Sect. 6.1 (see, for example, Bongaarts, 2001; Horrell & Krishnan, 2008), the data used in this chapter also shows that the size of *de-jure* FHHs is consistently smaller than that of both MHHs and *de-facto* FHHs. We observe a higher dependency ratio in *de-facto* FHHs compared to *de-jure* FHHs and MHHs. This observation is in line with the characteristics of FHHs found in the literature (Buvinić & Gupta, 1997; Kossoudji & Mueller, 1983; Quisumbing et al., 2001). Food shares of *de-jure* FHHs are slightly higher than those of *de-facto* FHHs and MHHs.

As can be seen in Table 6.1, in line with the observations regarding the differences in household incomes (or expenditure) by gender of the head of the household highlighted in the literature (see, for example, Buvinić & Gupta, 1997; Lampietti & Stalker, 2000), there are obvious differences in mean expenditure between the *de-jure* FHHs and MHHs categories. The mean expenditure of *de-jure* FHHs is consistently lower than those of the MHHs and *de-facto* FHHs. Figure 6.1 graphically represents the differences in mean household expenditure by the gender of the household head. Similar findings were noted in other studies in the Sri Lankan context (for example, see Aturupane et al., 1997; Jayathilaka, 2007; Jayasinghe, 2019).

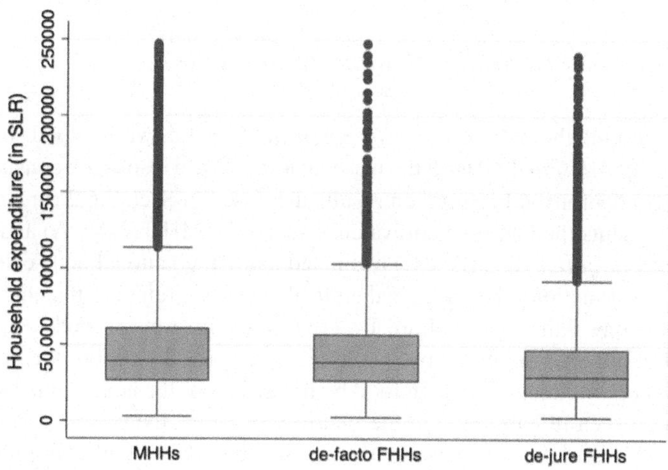

Fig. 6.1 Household expenditure by gender of the household head. *Source* Author's compilation based on Department of Census and Statistics (2017)

6.3.1.1 Methodology

The empirical analysis of this chapter is based on expenditure-dependent Engel equivalence scales, discussed in Sect. 2.3.3. The Engel curves for food were estimated separately for MHHs and *de-facto* and *de-jure* FHHs at the national and sectoral level, using the general form regression specification presented in Chap. 2 and reproduced here in Eq. (6.1).

$$w_f = \beta_0 + \beta_1 \ln \frac{x}{n} + \beta_2 \left(\ln \frac{x}{n}\right)^2 + \gamma_1 n_a + \gamma_2 n_c + \gamma_3 n_a n_c \quad (6.1)$$

The Engel equivalence scales were then estimated at the mean expenditure at the national and sectoral level.

The differences in poverty status between MHHs and *de-facto* and *de-jure* FHHs were investigated by looking at the change in percentage of households who fall below the official poverty line. For this purpose, the percentage of households that fall below the official poverty line in 2016 (SLR. 4,166 real total expenditure per person), under both the PCE and the EPEA approaches (as discussed in Sect. 5.5.3, in Chap. 5), were estimated and compared.

6.4 Results and Discussion

6.4.1 Food Consumption Economies of Scale

Table 6.2 presents the parameter estimates of the Engel curve for food, for MHHs and *de-facto* and *de-jure* FHHs, at the national level. These results are consistent with the Engel curves for food estimation results discussed in Sect. 5.6.2 (in Chapter 5). Table 6.3 presents the marginal equivalence scales for MHHs (Part A) and *de-facto* (Part B), and *de-jure* FHHs (Part C) estimated using the national and sectoral level Engle curve estimations. The marginal equivalence scale refers to the change in the cost of remaining at the same welfare level (measured in terms of food consumption) when household size increases by one adult. These results enable the analysis of consumption economies of scale in food by the gender of the head of the household.

As can be seen in Table 6.3, in the majority of cases, the marginal equivalence scales of *de-jure* FHHs are consistently smaller than those of the *de-facto* FHHs, while MHHs reflect the highest marginal equivalence scales. For example, at the national level, the marginal equivalence scale of 0.19 for the (2,0)–(3,0) household size of *de-jure* FHHs implies that when household size increases from two-adult (2,0) to three-adult (3,0), the three-adult household headed by the *de-jure* female head spends 19% more on food to remain in the same level of welfare (measured in terms of the budget share on food) of a two-adult *de-jure* FHH, compared to 21% of additional cost incurred by *de-facto* FHHs and 23% of additional cost incurred by

Table 6.2 Engel curves for food by gender of the head of the household

Variable (1)	Male-headed (2)	*De-facto*female-headed (3)	*De-jure*female-headed (4)
Constant (β_0)	1.560* (0.000)	4.532* (0.000)	6.561* (0.009)
Log of PCE (β_1)	−0.015* (0.000)	−0.354 (0.181)	−0.303* (0.015)
Square of Log of PCE (β_2)	−0.009 (0.963)	0.007 (0.123)	0.016* (0.019)
Number of adults (γ_1)	−0.041* (0.000)	−0.031* (0.000)	−0.040* (0.000)
Number of children (γ_2)	−0.060* (0.000)	−0.035* (0.000)	−0.048* (0.000)
Interaction term—number of children and adults (γ_3)	0.010** (0.079)	0.024* (0.0014)	0.011* (0.000)
R-squared	0.592	0.487	0.382

Note Dependent variable is expenditure share on food in each type of household head. *p*-values are given in parenthesis. * $P < 0.05$
Source Author's calculations based on Department of Census and Statistics (2017)

Table 6.3 Marginal equivalence scales for food by the gender of the head of the household

Household size(1)	National (2)	Urban(3)	Rural (4)	Estate(5)
Part A: *male-headed households*				
1,0–2,0	0.38	0.40	0.38	0.42
2,0–3,0	0.23	0.26	0.24	0.28
3,0–4,0	0.17	0.16	0.14	0.15
4,0–5,0	0.09	0.08	0.07	0.07
Part B: *de-facto female-headed households*				
1,0–2,0	0.35	0.40	0.38	0.43
2,0–3,0	0.21	0.25	0.24	0.24
3,0–4,0	0.16	0.14	0.12	0.13
4,0–5,0	0.09	0.07	0.06	0.08
Part C: *de-jure female-headed households*				
1,0–2,0	0.35	0.39	0.36	0.40
2,0–3,0	0.19	0.22	0.18	0.21
3,0–4,0	0.08	0.11	0.08	0.10
4,0–5,0	0.04	0.06	0.01	0.03

Source Author's compilation based on Department of Census and Statistics (2017)

MHHs. This suggests that *de-jure* FHHs materialise the highest level of economies of scale in food consumption (and hence experience lower marginal cost for an additional adult), followed by *de-facto* FHHs and MHHs at the national level as well as at the sectoral level.

This observation is not surprising as some research has suggested that low-income households tend to materialise greater economies of scale in consumption (see, for example, Jayasinghe et al., 2016; Jayasinghe & Smith, 2021; Leibtag & Kaufman, 2003) and the mean household expenditure data presented in Table 6.1 and Fig. 6.1 indicated that *de-jure* FHHs have the lowest household expenditure levels. The lower income (expenditure) levels in *de-jure* FHHs may lead them to economise their food consumption in various forms, such as the consumption of home-grown food and home-cooked food (Jayasinghe et al., 2017; Vernon, 2005), bulk-purchasing of food items and use of traditional food preservation methods (Kalansooriya & Chandrakumara, 2014; Leibtag & Kaufman, 2003). A study on Sri Lanka conducted by Kalansooriya and Chandrakumara (2014) revealed that the majority of the women in their sample used some kind of food preserving method to ensure the availability of food during scarce times. Kalansooriya and Chandrakumara (2014) also found that about 84% of the female household heads, in rural areas paid greater attention to preparing every meal at home for their family members to avoid the consumption of food prepared outside. Such practices among the female household heads, particularly in the face of lower-income levels and declining or no support from a male partner, may contribute towards the higher food consumption economies of scale in FHHs.

At the sectoral level, households belonging to all types of household heads in the rural sector are found to have the lowest marginal equivalence scales, hence the highest level of food consumption economies of scale. Due to the agricultural occupation base of households in this sector, they may consume more home-grown food than households in the other two sectors, leading them to achieve greater economies of scale. The marginal cost of an adult in the estate sector is either similar to that of the urban sector, or in some cases even higher than the urban sector, indicating that households in the estate sector materialise the least economies of scale in food consumption. When considering the estate sector, the majority of the population in the estate sector work in tea plantations and they live in line-housing provided by their employers. These conditions may generate time constraints and limited access to land for home production of food. Estate households experience a significantly lower standard of living with significantly different consumption patterns than the urban and rural sectors (Department of Census and Statistics, 2017). As such, they may also have limited access to electricity and household assets that may enable them to experience greater economies of scale compared to urban and rural households. According to Jayasinghe et al. (2018) estate sector households spent only about 1% of their total monthly expenditure on electrical appliances in 2012/2013. Therefore, as Beatty (2010) suggested, the estate sector households may be too poor to materialise economies of scale in consumption.

Table 6.4 Percentage of poor households by gender of the head of the household

Percentage of poor households (1)	National (2) (%)	Urban(3) (%)	Rural(4) (%)	Estate(5) (%)
Part A: *male-headed households*				
PCE	4.7	1.5	5.15	10.26
EPEA	3.5	0.5	3.1	6.2
Percentage decline	*1.2*	*1.0*	*2.0*	*4.1*
Part B: *de-facto female-headed households*				
PCE	7.2	3.1	6.8	20.0
EPEA	3.8	0.7	3.9	7.3
Percentage decline	*3.4*	*2.4*	*2.9*	*12.7*
Part C: *de-jure female-headed households*				
PCE	10.2	5.1	9.5	25.1
EPEA	6.1	2.1	4.9	10.0
Percentage decline	*4.1*	*3.0*	*4.6*	*15.1*

Note The official poverty line for 2016 is SLR. 4,166 real total expenditure per person
Source Author's compilation based on Department of Census and Statistics (2017)

6.4.2 *Implications on Poverty Measurements*

Table 6.4 provides the percentage of poor households by the gender of head of household under the conventional PCE and EPEA. Under PCE, the percentage of poor de-jure FHHs is greater than that of de-facto FHHs and MHHs at the national level and the sectoral level. The estate sector followed by the rural sector demonstrates the highest poverty rates in the country under the PCE method.

When EPEA, which takes the economies of scale in food consumption into account, is used, the poverty levels in both MHHs and FHHs decline significantly. This decline is larger among the FHHs, particularly among the *de-jure* FHHs in the majority of the cases, as shown in Table 6.4. For example, in 2016, at the national level, the difference between PCE and EPEA poverty figures among MHHs is 1.18%, while that of *de-facto* and *de-jure* FHHs is 3.41 and 4.13%, respectively. Nevertheless, even after economies of scale in food consumption is adjusted for, both types of FHHs show a higher percentage of poor households and that of *de-jure* FHHs are even higher than that of *de-facto* FHHs. For example, in the 2016 period, at the national level, the percentage of poor *de-facto* and *de-jure* FHHs is 3.80 and 6.05%, respectively, compared to 3.52% of poor MHHs under EPEA. The estate sector demonstrates the highest poverty levels under the EPEA as well.

6.5 Concluding Remarks

This chapter investigated whether the gender of the head of the household affects the level of food consumption economies of scale achieved by households and examined the effect of such differences in food consumption economies of scale on poverty measurements. The results revealed that *de-jure* FHHs materialise considerably higher levels of food consumption economies of scale than MHHs and *de-facto* FHHs. Overall, the poverty levels in MHHs and *de-facto* and *de-jure* FHHs decline after necessary adjustments for food consumption economies of scale are allowed for. However, the results revealed that, despite the high level of economies of scale in food consumption observed in *de-jure* FHHs, after adjusting poverty figures to incorporate such economies of scale *de-jure* FHHs still show the highest level of poverty in Sri Lanka. The findings of this study highlight the need for gender-sensitive development policies targeted towards poverty reduction and empowerment of disadvantaged women.

Acknowledgements Some of the materials in this chapter are reprinted by permission from the Springer Nature: Springer, Social Indicators Research, Poverty Implications of Household Headship and Food Consumption Economies of Scales: A Case Study from Sri Lanka, Jayasinghe, Maneka and Smith, Christine, Vol. 155, No. 1, 2021.

References

Amirthalingam, K., & Lakshman, R. (2013). Impact of displacement on women and female-headed households: A mixed method analysis with a microeconomic touch. *Journal of Refugee Studies, 26*(1), 26–46. https://doi.org/10.1093/jrs/fes007

Aturupane, H., Rodrigo, C. and Perera, S. (1997). Poverty among female-headed households in Sri Lanka. In *Reported prepared for the World Bank, Washington, DC, and the Regional Development Division, Ministry of Plan Implementation, Ethnic Affairs and National Integration.*

Balli, F., & Tiezzi, S. (2013). Declining equivalence scales and cost of children: Evidence and implications for inequality measurement. *The B.E. Journal of Economic Analysis & Policy, 13*(2), 761–782. https://doi.org/10.1007/s11150-009-9085-2

Beatty, T. K. M. (2010). Do the poor pay more for food? evidence from the United Kingdom. *American Journal of Agricultural Economics, 92*(3), 608–621.

Bongaarts, J. (2001). Household size and composition in the developing world in the 1990s. *Population Studies—A Journal of Demography, 55*(3), 263–279. https://doi.org/10.1080/00324720127697

Burch, T. K., & Matthews, B. J. (1987). Household formation in developed societies. *Population and Development Review, 13*(3), 95–511.

Buvinić, M., & Gupta, G. R. (1997). Female-headed households and female-maintained families: Are they worth targeting to reduce poverty in developing countries? *Economic Development and Cultural Change, 45*(2), 259–280. https://doi.org/10.1086/452273

Chant, S. (1997). Women-headed households: Poorest of the poor? Perspectives from Mexico, Costa Rica and the Philippines. *IDS Bulletin, 28*(3), 26–48.

Chant, S. (2004). Dangerous equations? how female-headed households became the poorest of the poor: Causes, consequences and cautions. *IDS Bulletin, 35*(4), 19–26.

Chant, S. (2011). *The international handbook of gender and poverty: Concepts, research, policy*. Edward Elgar.

Chant, S., & Campling, J. (1997). *Women-headed households: Diversity and dynamics in the developing world*. Palgrave Macmillan.

Chant, S. (2003). Female household headship and the feminisation of povery: Facts, fictions and forward strategies. *New Working Paper Series* (Vol. Issue 9). LSE Gender Institute.

De Silva, W. I. (2005). Family transition in South Asia: Provision of social services and social protection. *Asia-Pacific Population Journal, 20*(2), 13–46.

Deaton, A., & Paxson, C. (1998). Economies of scale, household size, and the demand for food. *The Journal of Political Economy, 106*(5), 897–930.

Department of Census and Statistics (DCS). (2017). *Household income and expenditure survey—2016*. Department of Census and Statistics.

Donaldson, D., & Pendakur, K. (2003). Equivalent-expenditure functions and expenditure-dependent equivalence scales. *Journal of Public Economics, 88*, 175–208.

Drèze, J., & Srinivasan, P. V. (1997). Widowhood and poverty in rural India: Some inferences from household survey dta. *Journal of Development Economics, 54*(2), 217–234. https://doi.org/10.1016/S0304-3878(97)00041-2

Greenstein, T. N. (2000). Economic dependence, gender, and the division of labor in the home: A replication and extension. *Journal of Marriage and Family, 62*(2), 322–335. https://doi.org/10.1111/j.1741-3737.2000.00322.x

Griffith, R., Leibtag, E., Leicester, A., & Nevo, A. (2009). Consumer shopping behavior: How much do consumers save? *The Journal of Economic Perspectives, 23*(2), 99–120.

Gunatilaka, R., & Vithanagama, R. (2018). Women's labour force participation in Sri Lanka's North. In *GrOW Working Paper Series*. McGill University and the International Development Research Centre.

Haddad, L., Peña, C., Nishida, C., Quisumbing, A., & Slack, A. (1996). Food security and nutrition implications of intrahousehold bias: A review of literature. *FCND Discussion Paper Series* (Vol. 19). International Food Policy Research Institute.

Handa, S. (1994). Gender, headship and intrahousehold resource-allocation. *World Development, 22*(10), 1535–1547. https://doi.org/10.1016/0305-750x(94)90036-1

Horrell, S., & Krishnan, P. (2008). Poverty and productivity in female-headed households in Zimbabwe. *The Journal of Development Studies, 43*(8), 1351–1380.

Ibnouf, F. O. (2012). The value of women's indigenous knowledge in food processing and preservation for achieving household food security in rural Sudan. *Journal of Food Research, 1*(1), 238–253. https://doi.org/10.5539/jfr.v1n1p238

Jayasinghe, M. (2019). Dynamic trends in household poverty and inequality in Sri Lanka: Do gender and ethnicity matter? *Journal of the Asia Pacific Economy, 24*(2), 208–223.

Jayasinghe, M., & Smith, C. (2021). Poverty implications of househild headship and food consumption economies of scale: A case study from Sri Lanka. *Social Indicators Research, 155*, 157–185.

Jayasinghe, M., Smith, C., Chai, A., & Ratnasiri, S. (2016). The implications of income dependent equivalence scales for measuring poverty in Sri Lanka. *International Journal of Social Economics, 43*(12), 1300–1314.

Jayasinghe, M., Chai, A., Ratnasiri, S., & Smith, C. (2017). The power of the vegetable patch: How home-grown food helps large rural households achieve economies of scale and escapep poverty. *Food Policy, 73*, 62–74.

Jayasinghe, M., Ratnasiri, S., Smith, C., & Chai, A. (2018). Domestic technology, consumption economies of scale and poverty: Evidence from Sri Lanka. *Applied Economics, 50*(16), 1777–1789.

Jayathilaka, R. (2007). Economic and demographic characteristics of poor female-headed households in Sri Lanka. *Sri Lanka Journal of Population Studies, 9*, 1–22.

Kalansooriya, C. W., & Chandrakumara, D. P. S. (2014). Women's role in household food security in rural Sri Lanka. *International Journal of Multidisciplinary Studies, 1*(1), 41–54.

Kennedy, E., & Haddad, L. (1994). Are pre-schoolers from female-headed households less malnourished? A comparative analysis of results from Ghana and Kenya. *Journal of Development Studies, 30*(3), 680–695. https://doi.org/10.1080/00220389408422332

Kennedy, E., & Peters, P. (1992). Household food security and child nutrition: The interaction of income and gender of household head. *World Development, 20*(8), 1077–1085. https://doi.org/10.1016/0305-750X(92)90001-C

Kossoudji, S., & Mueller, E. (1983). The economic and demographic status of female-headed households in rural Botswana. *Economic Development and Cultural Change, 31*(4), 831–859. https://doi.org/10.1086/451360

Koulovatianosa, C., Schroder, C., & Schmidt, U. (2005). On the income dependence of equivalence scales. *Journal of Public Economics, 89*, 967–996.

Krishnan, P., & Horrell, S. (2007). Poverty and productivity in female-headed households in Zimbabwe. *The Journal of Development Studies, 43*(8), 1351–1380. https://doi.org/10.1080/00220380701611477

Lakshman, R. W., & Jayasinghe, M. (2011). Livelihood development in a gender perspective—A case study from Sri Lanka. *Sri Lanka Economic Journal, 12*(2), 82–116.

Lampietti, J. A., & Stalker, L. (2000). Consumption expenditure and female poverty: A review of the evidence. *Policy Research Report on Gender and Development Working Paper Series 11*. World Bank.

Lanjouw, P., & Ravallion, M. (1995). Poverty and household size. *The Economic Journal, 105*(433), 1415–1434.

Leibtag, E. S., & Kaufman, P. R. (2003). *Exploring food purchase behavior of low-income households: How do they economize?* Agriculture Information Bulletin.

Logan, T. D. (2011). Economies of scale in the household: Puzzles and patterns from the American past. *Economic Inquiry, 49*(4), 1008–1028.

Mahadevan, R., & Jayasinghe, M. (2020). Examining multidimensional poverty in Sri Lanka: Transitioning through post war conflict. *Social Indicators Research, 149*, 15–39. https://doi.org/10.1007/s11205-019-02229-2

Quisumbing, A. R., Haddad, L., & Peña, C. (2001). Are women overrepresented among the poor? An analysis of poverty in 10 developing countries. *Journal of Development Economics, 66*(1), 225–269. https://doi.org/10.1016/S0304-3878(01)00152-3

Ramnarain, S. (2016). Unpacking widow headship and agency in post-conflict Nepal. *Feminist Economics, 22*(1), 80–105.

Ruwanpura, K. (2006). Conflict and survival. *Asian Population Studies, 2*(2), 187–200.

Ruwanpura, K. N., & Humphries, J. (2004). Mundane heroines: Conflict, ethnicity, gender, and female headship in Eastern Sri Lanka. *Feminist Economics, 10*(2), 173–205+267. https://doi.org/10.1080/1354570042000217766

Samarasinghe, V. (1993). Access of female plantation workers in Sri Lanka to basic-needs provision. In J. Momsen & V. Kinnaird (Eds.), *Different places, different voices: Gender and development in Africa, Asia, and Latin America* (pp. 131–145). Routledge.

Van de Walle, D. (2013). Lasting welfare effects of widowhood in Mali. *World Development, 51*, 1–19. https://doi.org/10.1016/j.worlddev.2013.05.005

Van der Lippe, T., & Siegers, J. J. (1994). Division of household and paid labour between partners—Effects of relative wage rates and social norms. *International Review of Social Sciences, 47*(1), 109–136. https://doi.org/10.1111/j.1467-6435.1994.tb02252.x

Vasudevan, R. (2013). *Everyday resistance: Female-headed households in Northern Sri Lanka*. Graduate Institute Publications.

Vernon, V. (2005). *Food expenditure, food preparation time and household economies of scale*. State University of New York.

Visaria, P. (1980). Poverty and living standards in Asia. *Population and Development Review, 6*(2), 189–223.

Chapter 7
Conclusion and Policy Implications

7.1 Introduction

Reducing poverty, eradicating hunger and improving the wellbeing of all people are key development priorities under the United Nations SDGs that are intended to be achieved globally by the year 2030. The analysis undertaken in this book directly contributes to this global agenda by enhancing the understanding of poverty levels through accurate measurement of poverty and thereby enabling the identification of the poorest of the poor. The empirical analysis which focuses on the household food consumption patterns and the drivers of food consumption economies of scale also contribute to the knowledge of enhancing food security and thereby eradicating hunger of the poor and vulnerable groups in society.

Throughout this book, we argue that the consumption economies of scale play a significant role in accurately identifying the poorest of the poor. This is because the per capita based poverty measures, which are commonly used to measure poverty, ignore the presence of consumption economies of scale in large households, leading to an underestimation of the welfare in large households. The use of household size-specific equivalence scales is regarded as the most appropriate way to adjust household income for differences in household size and composition (Deaton & Muellbauer, 1980). While these equivalence scales can be estimated using very complicated econometric methods, policymakers in the real world often use simple and predefined equivalence scales such as the OECD-modified scales. These OECD-modified scales assign 1 to the household head, 0.5 to each additional adult and 0.3 to each child. However, these simple scales ignore some other important socio-demographic characteristics and consumption behavioural choices that may significantly influence the level of consumption economies of scale achieved by a particular household.

This book explored how the magnitude of the consumption economies of scale in food achieved by households was influenced by their socio-demographic characteristics and behavioural choices. For this purpose, household specific equivalence scales that take all these aspects into account were estimated using the Engle equivalence scales approach. The empirical analysis also investigates the impact of such

M. Jayasinghe, *Poverty, Food Consumption, and Economic Development*,
https://doi.org/10.1007/978-981-16-8743-3_7

economies of scale on the poverty measurements, enabling the identification of the poorest of the poor. Such poverty rates can be used to assess the effectiveness of the poverty reduction strategies that have been adopted in a country, as well as to redirect the existing poverty reduction strategies to better target the neediest segments in the population and thereby achieve poverty reduction targets. Sri Lanka was used as a case study for the empirical analysis in this book. This book focused on the economies of scale arising from food consumption because the expenditure on food represents a major proportion of household expenditure in developing countries and food security is a major challenge in many developing countries, including Sri Lanka.

The literature review highlighted three key knowledge gaps regarding consumption economies of scale, particularly those of food consumption (see Chap. 1):

- The absence of any attempt in the literature to investigate the role of home-grown food consumption in enhancing economies of scale in food consumption and the lack of research on the implications of home-grown food on food security, even though home-grown food makes a substantial contribution to the food supply of rural households in many developing countries.
- The absence of any attempt in the literature in exploring the role of increased household electrification and domestic technology adoption in enhancing food consumption economies of scale and the lack of clearly identified pathways on the ways in which electricity access enhances household welfare.
- Little attention is being given to exploring the relationship between food consumption economies of scale and the gender of the household head, even though the association between gender and poverty has been extensively discussed in the literature.

This overall lack of attention to the effects of various household attributes on economies of scale in food consumption prevents the formation of a more realistic picture of the status of poverty and its underlying causes. This book has addressed this issue through a rigorous empirical investigation of how these household characteristics and behavioural choices (such as consumption of home-grown food, and domestic technology adoption) affect economies of scale in food consumption and poverty. The empirical analysis of this book is based on the latest available (2016) Household Income and Expenditure Survey (HIES) of Sri Lanka conducted by the Department of Census and Statistics (2017).

This book developed two empirically testable theoretical models; one in Chap. 4 to investigate the presence of economies of scale in the consumption of home-grown food and the other one in Chap. 5 to examine the role of domestic technology adoption in economies of scale in food consumption. The empirical analysis in this book was based on the Engel equivalence scales method (in Chaps. 5 and 6) and a number of multiple regression models (in Chaps. 4 and 5). The empirical evidence presented in this book makes a significant contribution to the literature on economies of scale in consumption, food security and poverty measurements particularly in the context of developing countries. The policy implications that arise from these research findings, both for effective targeting of the poorest of the poor and for uplifting their standard of living in developing countries, represent another important contribution emerging from this book.

The purpose of this final chapter is threefold: firstly, to summarise the main empirical findings of the study; secondly, to discuss the policy implications of these findings and present some policy suggestions to improve the standard of living of vulnerable population sub-groups in Sri Lanka and other developing countries with similar socio-economic backgrounds (Sect. 7.2), and; thirdly, to briefly discuss the limitations of this study and propose some insights for future research (Sect. 7.3).

7.2 Summary of the Main Findings and Policy Implications

Chapters 1 and 2 of this book provided the background and methodology for the research carried out in this book. As presented in Chap. 1, the central research question this book seeks to answer is *How do the households' characteristics and behavioural choices affect the level of food consumption economies of scale enjoyed by households and what are the implications of these food consumption economies of scale on household poverty status?*

In the process of answering this central research question, this book specified four additional research questions that have both academic and policy relevance. These include:

R.Q.1. How does the household choice to consume home-grown food affect food consumption economies of scale and enhance food security?

R.Q.2. How does domestic technology adoption and household electricity access affect food consumption economies of scale?

R.Q.3. Do food consumption economies of scale vary by the gender of the head of the household?

R.Q.4. How do the food consumption economies of scale affect the poverty measurements?

Chapter 2 presented a detailed analysis of consumption economies of scale and the main methodological approach adopted in this book to capture these economies of scale in consumption. An in-depth explanation of the data used in this study was given in Chap. 3. Chaps. 4 to 6 were presented as a series of interconnected empirical chapters that seek answers for each of these four research sub-questions. This section summarises the key findings in these empirical chapters and key policy implications emerging from these results.

7.2.1 *How Does the Household Choice to Consume Home-Grown Food Affect Food Consumption Economies of Scale and Enhance Food Security? (R.Q.1)*

Chapter 4 presented a theoretical model, followed by an empirical application, that investigates the presence of higher food consumption economies of scale in home-grown food than in purchased food. The results revealed that the household size elasticity of home-grown food and purchased food is negative, implying that there are food consumption economies of scale in both types of food. However, the magnitude of the household size elasticity in home-grown food is higher than that of purchased food, suggesting that home-grown food provides higher food consumption economies of scale than market-purchased food. The empirical results also revealed that rural households and low-income households are more inclined to fulfil their food needs by consuming home-grown food than their urban and high-income counterparts, respectively. Chapter 4 also investigated whether the reduction in per capita food expenditure when large households consume more home-grown food is due to the realisation of economies of scale in food consumption or due to a reduction in overall food consumption. The results revealed that consumption of home-grown food is, in fact, associated with about 18% higher per capita calorie consumption, and does not adversely affect the per capita calorie intake of large households.

Drawing on these observations, an important policy implication arising from this study is to encourage greater use of the 'backyard' food production among low-income households. The food sourced from home gardens may enhance the availability of and accessibility to affordable food, ensuring food security for disadvantaged households. In this context, the relevant government organisations and authorities can support backyard vegetable gardening via information campaigns on educating households about the types of food crops that can be cultivated in small plots of land (FAO, 2011). For those households which do not have access to their own plots, the provision of small alternatives such as containers, trellises and hydroponic gardening equipment could also be considered. Additionally, a local adaptation of the concept of 'community gardening', where a plot of private or public land is used for gardening by a group of people, is also a viable option for communities such as low-income and less privileged households in the estate sector which do not have adequate land to grow their own food. TheFAO (1997) has suggested other ways through which policymakers can promote sustainable backyard gardening in developing countries such as the promotion of community seed-sharing projects and the distribution of compost bins which, when accompanied by the provision of training and information required to use them, may provide a cheap and sustainable source of fertilizer. While this source of food may not be as relevant in a developed country context with high levels of urbanisation, and where households possess high opportunity costs of time, it is highly relevant for many developing countries that still have a large proportion of their population situated in rural areas, many of whom work in the agricultural sector and own small plots of land.

7.2.2 How Does the Domestic Technology Adoption and Household Electricity Access Affect Food Consumption Economies of Scale? (R.Q.2)

Chapter 5 examined the effects of domestic technology adoption and access to electricity on food consumption scale economies in Sri Lanka. Firstly, a theoretical model was developed to identify how domestic technology adoption may enable households to achieve higher economies of scale in food consumption. Secondly, the validity of this model was tested using regression analysis. In the actual estimation, the expenditure on electricity was used as a proxy for expenditure share on domestic technology. Thirdly, using an alternative approach, Engel equivalence scales, the empirical analysis of the chapter examined whether households with access to electricity are able to reap more benefits of economies of scale in food consumption than the households who do not have access to electricity. The presence of such higher food consumption economies of scale is reflected by lower equivalence scales (hence lower marginal cost) associated with households with access to electricity.

Empirical results revealed that higher expenditure shares on technology are associated with lower PCE on food. This is because the adoption of domestic technology, facilitated by electrification, appears to reduce the food preparation, processing, and preservation costs. Furthermore, the reduction in time required to perform food preparation related activities, with the use of domestic technology, may have encouraged time scarce households to prepare food at home rather than having meals prepared outside. This may have enabled these households to reap the benefits of consumption economies of scale through increasing returns to scale in home production. The estimated Engel equivalence scales revealed that households with access to electricity report lower equivalences scales than households who do not have access to electricity. The results also indicated that a significant number of households who have been categorised as poor, based on per capita measures, appear to move out of poverty if food consumption economies of scale realised through domestic technology adoption are allowed for. The novelty of this study was to link technology adoption and access to electricity to the analysis of economies of scale from the consumption aspect, and thereby to incorporate the economies of scale effect of domestic technology adoption in the measurement of poverty.

The role of improved access to efficient energy sources and technology adoption has received significant attention in augmenting the welfare of the population. Facilitating access to efficient energy sources has been widely applied as a part of the development agenda in many developing countries. The findings of this study further reinforce the importance of household electrification and the adoption of an appropriate level of technology in everyday life. In terms of policy implications, these results highlight the necessity for policies that facilitate household electrification and encourage enhanced levels of domestic technology adoption. Provision of subsidies that allow for a greater level of access to electricity in rural areas and subsidising the electricity prices for low-income households in disadvantaged areas within the rural and estate sectors, for example, could act to improve the standard of living of these households in a well-targeted manner (Global Energy Assessment, 2012). Policies directed towards increased use of solar power to generate electricity such as sharing relevant information on the benefits of solar power, introducing rebates for households to install solar panels, and subsidising the installation of solar panels for very poor households, may be considered. This may be especially useful in those rural areas where access to the grid is prohibitively expensive. Furthermore, allocation of funding, or provision of interest-free loans for low-income households to purchase energy-saving domestic appliances, may also be a valuable option to consider. This may not only make their transition from traditional methods to technologically advanced and efficient methods (that enable greater food consumption scale economies) both smoother and financially less stressful, it may also ease the burden of their electricity bills in the longer term.

7.2.3 Do Food Consumption Economies of Scale Vary by the Gender of the Head of the Household? (R.Q.3)

Chapter 6 investigated how the gender of the head of the household affects the household ability to reap the benefits of food consumption economies of scale. The analysis was based on Engel equivalence scales estimated separately for MHHs and *de-facto* and *de-jure* FHHs. The impact of these differences in food consumption economies of scale, based on the gender of the household headship, on the poverty status of households headed by males and females was also examined in this chapter.

The results revealed that *de-jure* FHHs materialise considerably higher levels of food consumption economies of scale than MHHS and *de-facto* FHHS. The results also revealed that despite the high level of economies of scale in food consumption observed in *de-jure* FHHs, after adjusting poverty figures to incorporate such economies of scale, they still show the highest level of poverty in Sri Lanka. To this end, what our results really highlight is that *de-jure* FHHs overrepresent the poor in Sri Lanka. Consequently, the findings from this study lead to some important policy implications.

The results suggest the need for government interventions aimed at removing obstacles for female labour force participation. Increased participation in paid work is expected to raise income levels of women and female household heads, and thereby drive them out of poverty. A common barrier that leads women to restrain from engaging in paid employment relates to time, skills and capital (World Bank, 2013). Access to quality and affordable childcare services has proven effective in enhancing the labour supply of mothers with preschool-aged children (Lefebvre & Merrigan, 2008). Additionally, improving the quality and availability of utility-based services (such as electricity, water supply and sanitation) can reduce the time women spend on performing housework and thereby free up time for economic participation (World Bank, 2013). The electricity accessibility may also enhance the possibility of using various food management practices, such as bulk buying during the season and preserving for later use and home-gardening of basic food items enhancing the food security of vulnerable households (Jayasinghe & Smith, 2021; Jayasinghe et al., 2018).

7.2.4 How Do the Food Consumption Economies of Scale Affect the Poverty Measurements? (R.Q.4)

Overall, the findings of this book reinforce the importance of the use of expenditure per equivalent adult (EPEA) over per capita expenditure (PCE) when comparing household poverty status. In particular, the findings in Chaps. 5 and 6 indicated how the food consumption economies of scale influence the poverty measurements in Sri Lanka. The results in Chap. 5 revealed that the percentage of poor households declines considerably when food consumption economies of scale due to technology adoption, facilitated by electricity is taken into consideration, that is when EPEA is used instead of the PCE. Similarly, poverty estimations in Chap. 6 revealed that the percentage of poor households declines when economies of scale in food consumption based on the gender of the household head is incorporated in poverty measurement estimations. The poor households under the EPEA method are the poorest in the country, while those who moved out of poverty after allowing for food consumption economies of scale may be poor, but they are better off than those who are still poor even after adjusting for food consumption economies of scale. Poverty alleviation programs should first target these poorest households. However, it should be noted that even a small reduction in the level of consumption economies of scale, in households that just moved out of poverty after allowing for food consumption economies of scale, may lead these households to fall back to 'poor' status.

In general, the analysis in this book provided important policy insights for improving the standard of living of the poorest households in the community, with a special emphasis on policies focused on gender (MHHs, *de-facto* and *de-jure* FHHs) and sector (urban, rural, and estate) specific development policies. The findings of this study provide some insights on the possible ways by which the standard of living of vulnerable people can be improved, other than the provision of cash-handouts to supplement income. These include facilitating and encouraging them to grow some food for their own consumption and facilitating increased access to effective energy sources and appropriate domestic technology. These research findings also provide useful insights for international aid agencies to use when implementing their initiatives to reduce poverty and vulnerability and thereby achieve a higher standard of living.

7.3 Limitations and Future Research

The analysis in this book is not without limitations, particularly those related to (1) data deficiencies, (2) empirical methodology, and (3) empirical analysis. When considering the limitations associated with data deficiencies, firstly, the food consumption expenditure data in the dataset does not allow recognition of the differences in the quality of food consumed by various households. In some instances, the higher/lower marginal cost of food consumption may be associated with quality-related matters, in addition to the quantity consumed, which is, unfortunately, unable to be distinguished from available data. Secondly, the empirical analysis on the implications of domestic technology adoption and electricity access is purely based on household electricity expenditure data. The exact data on domestic technology usage and electricity usage would have enabled more accurate analysis of the link between consumption economies of scale and technology adoption.

Several limitations could be associated with the empirical methodology adopted in this book. Firstly, as the empirical results of this book are based mainly on the Engel equivalence scales method, the results may suffer from the common issues associated with the Engel method, as discussed in Chap. 2. Secondly, due to issues related to identification of consumption economies of scale and differences in needs and preferences of individuals in the household, in this book, it was assumed that all of the adults and children, respectively, in the households have identical tastes, irrespective of their age and gender. Furthermore, the analysis in this book did not incorporate the differences in household preferences of the rich, poor and very poor household categories as HIES data does not support that type of investigation. This is particularly relevant with respect to consumption of processed food, food away from home and the use of electrical appliances in food consumption related activities. Future studies should consider a relaxation of this assumption, using more advanced methodologies. Thirdly, this book investigated only the extent of and the factors that affect economies of scale in consumption arising from food consumption and its implications for the measurement of poverty. The analysis in this book did not attempt to explore the effects of household demographic, socio-economic and behavioural characteristics on the consumption economies of scale in non-food items, or the implications of this for the measurement of poverty. This part of the analysis represents an interesting avenue for future research, especially as the nation moves from a low to a middle-income country and so the expenditure share on food declines in line with international norms.

In terms of the limitations relating to the empirical analysis, the findings that the consumption of home-grown food generates greater economies of scale also requires further research, including replication of this method to home-grown data from other countries as well as testing the robustness of these results using primary data. To this end, some exciting avenues remain for extending the current study in future research on food consumption, food security and poverty measures.

References

Deaton, A., & Muellbauer, J. (1980). *Economics and consumer behavior*. Cambridge University Press.

Department of Census and Statistics (DCS). (2017). *Household income and expenditure survey—2016*. Department of Census and Statistics.

FAO. (1997). *Agriculture food and nutrition for Africa—A resource book for teachers of agriculture*. FAO.

FAO. (2011). *The state of food insecurity in the world*. Food and Agricultural Orgnization of the United Nations.

Global Energy Assessment. (2012). *Global energy assessment—Toward a sustainable future*. International Institute for Applied Systems Analysis.

Jayasinghe, M., & Smith, C. (2021). Poverty implications of househild headship and food consumption economies of scale: A case study from Sri Lanka. *Social Indicators Research, 155*, 157–185.

Jayasinghe, M., Ratnasiri, S., Smith, C., & Chai, A. (2018). Domestic technology, consumption economies of scale and poverty: Evidence from Sri Lanka. *Applied Economics, 50*(16), 1777–1789.

Lefebvre, P., & Merrigan, P. (2008). Child-care policy and the labor supply of mothers with young children: A natural experiment from Canada. *Journal of Labor Economics, 26*(3), 519–548.

World Bank. (2013). Low female labor force participation in Sri Lanka: Contributory factors, challenges and policy implications. In *South Asia: Human development sector discussion paper series*. World Bank.

Author Index

M. Jayasinghe, *Poverty, Food Consumption, and Economic Development*, https://doi.org/10.1007/978-981-16-8743-3

Subject Index

M. Jayasinghe, *Poverty, Food Consumption, and Economic Development*, https://doi.org/10.1007/978-981-16-8743-3

The manufacturer's authorised representative in the EU is Springer Nature Customer Service Centre GmbH, Europaplatz 3, 69115 Heidelberg, Germany. If you have any concerns regarding our products, please contact ProductSafety@springernature.com

Printed and bound by CPI Group (UK) Ltd, Croydon, CR0 4YY
30/09/2026
02225307-0005